Robert Dunsmuir, 1825-1889.

Lynne Bowen

Since she completed a Master of Arts degree in Western Canadian History at the University of Victoria in 1980, Lynne Bowen has pursued a career writing social history with a popular appeal. Her first book, *Boss Whistle* (1982), won the 1983 Eaton's B.C. Book Award. Her second book, *Three Dollar Dreams* (1987), received the Lieutenant-Governor's Medal for Writing British Columbia History and was shortlisted for the Roderick Haig-Brown B.C. Book Prize. *Muddling Through: The Remarkable Story of the Barr Colonists* (1992) won the Hubert Evans Non-Fiction B.C. Book Prize. Her books have also won the Canadian Historical Association Regional Certificate of Merit twice – once for British Columbia and once for the Prairies/Northwest Territories. Her most recent book is *Those Lake People: Stories of Cowichan Lake* (1995). In 1998 her story "The Crosscut" was included in *Winds Through Time, An Anthology of Historical Young Adult Fiction.*

Lynne Bowen lives in Nanaimo, on Vancouver Island, where she and her husband, Dick, spend their spare time cycling, kayaking, and reading. Since 1992 she has been the Maclean Hunter Co-Chair of Creative Non-fiction Writing at the University of British Columbia.

THE QUEST LIBRARY
is edited by
Rhonda Bailey

The Editorial Board is composed of
Ven Begamundré
Lynne Bowen
Janet Lunn

Editorial correspondence:
Rhonda Bailey, Editorial Director
XYZ Publishing
P.O. Box 250
Lantzville BC
V0R 2H0
E-mail: xyzed@bc.sympatico.ca

In the same collection

Betty Keller, *Pauline Johnson: First Aboriginal Voice of Canada*.
Dave Margoshes, *Tommy Douglas: Building the New Society*.
John Wilson, *Norman Bethune: A Life of Passionate Conviction*.
Rachel Wyatt, *Agnes Macphail: Champion of the Underdog*.

Robert Dunsmuir

Copyright © 1999 by Lynne Bowen and XYZ Publishing.
Second printing 2004

All rights reserved. The use of any part of this publication reproduced, transmitted in any form or by any means, electronic, mechanical, photocopying, recording, or otherwise, or stored in a retrieval system without the prior written consent of the publisher – or, in the case of photocopying or other reprographic copying, a licence from Canadian Copyright Licensing Agency – is an infringement of the copyright law.

Canadian Cataloguing in Publication Data

Bowen, Lynne, 1940-

    Robert Dunsmuir : laird of the mines

    (The Quest Library ; 2)
    Includes bibliographical references and index.
    ISBN 0-9683601-3-0

    1. Dunsmuir, Robert, 1825-1889. 2. Coal mines and mining – British Columbia – Vancouver Island – History – 19th Century. 3. Vancouver Island (B.C.) – Biography. 4. Businessmen – British Columbia – Vancouver Island – Biography. I. Title. II. Series.

FC3823.1.D86B68 1999   971.1'203'092   C99-941235-3
F1089.V3B68 1999

Legal Deposit: Fourth quarter 1999
National Library of Canada
Bibliothèque nationale du Québec

XYZ Publishing acknowledges the support of The Quest Library project by the Canadian Studies Program and the Book Publishing Industry Development Program (BPIDP) of the Department of Canadian Heritage. The opinions expressed do not necessarily reflect the views of the Government of Canada.

The publishers further acknowledge the financial support our publishing program receives from The Canada Council for the Arts, the ministère de la Culture et des Communications du Québec, and the Société de développement des entreprises culturelles.

Chronology: Lynne Bowen
Index: Lynne Bowen
Layout: Édiscript enr.
Cover design: Zirval Design
Cover illustration: Francine Auger

Printed and bound in Canada

XYZ Publishing   Distributed by: General Distribution Services
1781 Saint Hubert Street   325 Humber College Boulevard
Montreal, Quebec H2L 3Z1   Toronto, Ontario M9W 7C3
Tel: (514) 525-2170   Tel: (416) 213-1919
Fax: (514) 525-7537   Fax: (416) 213-1917
E-mail: xyzed@mlink.net   E-mail: customer.service@emailgw.genpub.com

LYNNE BOWEN

DUNSMUIR

Robert

LAIRD OF THE MINES

To Dani,
Here's to your love
of history

Lynne Bowen

Nanaimo, May 3rd, 2012

*This book is dedicated to
Thora Howell
Friend of the written word*

> A capitalist is a man who lives on less than he earns.
>
> – Robert Dunsmuir

Contents

1 A Better Class of People	1
2 The Wilds of Vancouver Island	17
3 So Useful a Man	31
4 For He's a Jolly Good Fellow	47
5 The Managing Partner	61
6 Bloodshed Among Us	77
7 Sunday Soldiers	91
8 "In Full Everyday Working Blast"	103
9 A Princely Fortune	117
10 Laird of the Mines	131
Epilogue	145
Chronology of Robert Dunsmuir (1825-1889)	149
Sources Consulted	169
Index	173

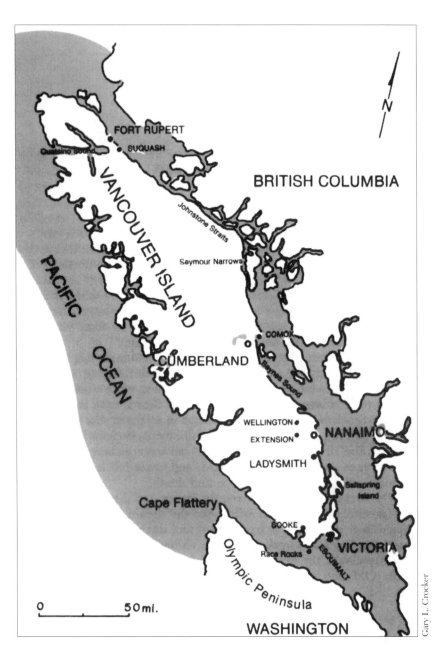

Coal Mining Communities of Nineteenth-Century Vancouver Island.

1

A Better Class of People

The letter is written in Greek and Spanish. The writer has used black ink and a single sheet of paper, which has been ripped from a pad and placed inside an envelope addressed in a large scrawling hand to the Hon. Robert Dunsmuir, Esquimalt and Nanaimo Railway, Victoria, British Columbia.

When Robert Dunsmuir opens the envelope on the third day of August 1888, the first thing he notices is the outline of a hand at the bottom of the page. Having no knowledge of Greek or Spanish, he asks his private secretary to bring a Spanish dictionary and the two men make a rough translation of the message: "You have to die very soon. By order. The Black Hand."

Robert Dunsmuir

The Black Hand is the favourite device of Spanish anarchists and New York City extortionists. It is so unlikely such sinister fellows are practising their illegal activities on Vancouver Island that Mr. Dunsmuir, who is about to attend a meeting, does nothing more than take the letter with him to show it to his friends.

But Robert Dunsmuir's friends are men in the highest positions in the province. The attorney general reads the offending letter and shows it to a judge. Then, before anyone can even offer an opinion, a second letter, also written in Greek and Spanish, arrives at Mr. Dunsmuir's office.

Victoria is not without its learned men in 1888. Maurice Lopatecki, Ph.D., has been in the city for only eight months, but he is already known as a classical Greek scholar and a former resident of Spain. His translation of the second letter persuades Mr. Dunsmuir to report the matter to the police:

> We send you another mandate. Beware. We know all that you are doing. Prepare for death. He who acts unjustly must be brought before the Judges to be punished.

No one threatens the province's wealthiest citizen twice in two days and gets away with it for long. Dunsmuir consults the police and tells them that he will pay for any additional expenses they might incur. Then, just in case they are unable to handle the situation, he engages Mahoney's Detective Agency of San Francisco and sends his secretary, William Whyte, on a private quest for information.

Laird of the Mines

Although Victoria is the provincial capital, it is not a very big place. Forty years ago, it was only a Hudson's Bay Company fort. In such a city, everyone knows everyone else and strangers are easily identified. Within hours, Whyte has a suspect. By comparing the handwriting in the letters to the handwriting on a prescription written in a drug store, Whyte is able to inform his employer that one G. H. Griffin, MD, is the likely culprit. Dunsmuir sets the San Francisco detective on the doctor's trail.

Since both the letters have been mailed at the same post office – the one on the corner of Pandora and Blanshard – Dunsmuir also pays for a postal clerk to sit near the drop box and wait for Griffin to mail another letter. When he sees the culprit drop the envelope in the slot, the clerk is to blow a police whistle to alert a deputy sheriff, who is hiding across the street. He will race over to make the arrest.

The two men stay at their posts for four days and do not see a single suspicious person mail a letter. The deputy sheriff, however, does see Dr. Griffin enter the post office through a door not usually open to the public. And during this time Mr. Dunsmuir does receive a third letter, which has been mailed at the very post office they have been watching so carefully. The letter says:

> On the 28th September we are going to kill you. It makes no difference if you have detectives. They are fools and cannot do anything. Prepare yourself. The time comes near when you will be no more.

Robert Dunsmuir

Robert Dunsmuir has lived for sixty-three years and he has endured much in that time. When he was seven years old, he and his sister were the only members of their family to survive a mysterious epidemic. At twenty-four he withstood the hardships of a six-month-long voyage from Scotland around Cape Horn. As an indentured coal miner, he was reviled by his fellow workers; as a mine owner, he was cursed by his employees. He has survived strikes and fires and explosions. No one gets the best of such a man, least of all a prankster calling himself The Black Hand.

The Black Hand's September 28th deadline comes and goes without incident. By October, Mr. Dunsmuir has received two more letters, and The Black Hand has broadened his list of victims to include Lopatecki and the San Francisco detective. But the large scrawling handwriting on all the letters eventually betrays Griffin, and he is arrested and taken to jail to await trial.

The trial is the most exciting thing that has happened in Victoria since the Marquis of Lorne and his wife, Queen Victoria's daughter, the Princess Louise, visited six years before. Crowds line the streets to witness the passage from the jail to the courthouse of the man who dared to threaten the illustrious Robert Dunsmuir. Gustavus Hamilton Griffin, late of London, Edinburgh, Paris, Athens, and Los Angeles, is a practitioner of medicine and surgery, a speaker of seven languages, a player of several musical instruments, a manufacturer of champagne, and, according to the bigotted language of the times, a hater of "Mr. Dunsmuir, niggers, Chinese, [and] Jews."

Laird of the Mines

Jaunty in a new tweed suit of the latest cut, the accused steps down from the carriage. He has a fresh flower in his buttonhole, a glossy silk hat on his head, and a toothpick held carelessly in his teeth. The hat and toothpick are gone, however, when he stands before Sir Matthew Baillie Begbie, first Chief Justice of British Columbia, a man known not only for his fairness but also for his lack of tolerance for those who threaten good order. Judge Begbie's address to the jury leaves no doubt what he thinks about Griffin's crime.

"When a person is threatened with death by a man face to face," Begbie intones, "he knows whom to protect himself against. [But] when he receives anonymous letters containing such threats he does not know whom to guard against. The offense is a very serious one against every member of society." He sentences Griffin to "penal servitude for five years."

Poor old Griffin. When he arrived in Victoria, a few steps ahead of an American bail bondsman who wanted to talk to him about his recent stealthy departure from Los Angeles, all he wanted to do was make a little quick money. When he heard that Robert Dunsmuir was building himself a castle and that he had made his money in coal mines in Nanaimo and Cumberland, Griffin bought some land near Nanaimo and tried to sell it to Dunsmuir on the chance that there was coal underneath it. Dunsmuir had rebuffed him. "I told him I didn't want to buy any more coal lands. I had enough."

As the holder of the Esquimalt and Nanaimo Land Grant, which included most of the accessible land on the east side of Vancouver Island from Victoria to Campbell River, Dunsmuir indeed had more than enough coal lands. But Griffin was outraged at being rebuffed. He told his clerk that Dunsmuir was "a very dangerous man, a son of a bitch." But Griffin had chosen the wrong man to threaten.

∞

Robert Dunsmuir was at the height of his powers in 1888. He had enormous wealth, political influence, a large, robust family, and two sons to follow in his footsteps. The crowning glory of his life was the stone castle that had been rising since the summer before on the highest hill overlooking Victoria.

Any Scot worthy of his heritage would build his house of stone if he could. All the important buildings in Scotland are built of stone, even the Presbyterian church in Ricarton near Kilmarnock in Ayrshire. Like Dunsmuir's castle, the church sits high on a hill. The tower over its main door climbs in wedding-cake tiers to a bell chamber that overlooks a small graveyard where careful rows of tombstones glow with a thin patina of moss.

Here are two headstones with the surname "Dunsmore" chiselled into each of their mottled surfaces. In the days when many people were illiterate and names were often spelled more than one way, Dunsmore was interchangeable with Dunsmuir. Beneath the two headstones lie eight Dunsmuirs: a husband and wife, their two daughters who died in infancy, their son and his wife and two daughters. Of the eight Dunsmuirs lying there, five of them died in the same month. The cause of their deaths – typhoid? cholera? – is nowhere recorded. Nor is the birth seven years earlier of the younger couple's surviving son, Robert.

In later years, when he was famous and people asked him about his origins, Robert Dunsmuir would say only that he was born in 1825 in Hurlford, which was a small village east of Kilmarnock. No birth certificate confirms this, however, nor is there any record of his baptism. His death certificate lists his birthplace as Burleith, not Hurlford. The actual location of Dunsmuir's birth is a small point – in the nineteenth century the villages and mines of this Scottish coal shire clustered closely together – but the lack of any record of his birth and his apparent reluctance to discuss it is strange. Some researchers, seeking to explain his sudden departure from Scotland as a young man, speculate that he was born before his parents were married and that he was ashamed of this.

Just fifty years before Robert's birth, the Emancipation Act of 1775 had ended slavery in the Scottish coal fields. Before that date, miners and their children had been bound as serfs to the coal they worked. But since emancipation they had become proud and prosperous, and some, like Robert's father and grandfather, had risen even higher. They had become coal masters or mine managers for the wealthy aristocrats who owned most of the coal fields in Scotland.

Pregnancy before marriage was not uncommon in nineteenth-century Scotland, but "antenuptial fornication," as the act of intercourse before marriage was called, was still considered a sin. Until the guilty man and woman confessed that sin to their congregation, they were barred from the church, and their children could not be baptized.

Whether or not Robert's parents were married when he was born, their deaths when he was seven years old are reason enough to explain why he didn't wish to speak about his beginnings. To relive that awful time in 1832, to remember his mother, Elizabeth, dying on a grim August day to be followed only five days later by his father, James, and six days after that by his grandmother, Jean, was reason enough to wish to forget the past. Then his two infant sisters, Marion and Elizabeth, died before the end of the month, leaving their widowed grandfather to raise Robert and his four-year-old sister, Jean.

Despite the enormous tragedy, Robert's and Jean's lives with Grandfather Dunsmuir were comfortable and secure for a time. The older man had turned the profits from a coal yard he owned on the main street of Kilmarnock into the building of a fine new house. But three years later he too was dead, and though the old man left enough money to provide for the two children's schooling, they had to adjust once again to losing the person closest to them.

The grandson and son of coal masters was to be a coal master too. At ten years old, Robert enrolled in the Kilmarnock Academy just down the street from his grandfather's coal yard. Later, he travelled to Paisley, twenty-five kilometres to the north, to learn the technical aspects of mining at the Mercantile and Mechanical School. Six years after his grandfather's death, he moved in with his father's sister, Jean, and her husband,

Boyd Gilmour, another coal master who began to teach him the more humble skills of the practical coal miner.

An early portrait of Robert Dunsmuir reveals a serious young man in a black suit complemented by a shiny cravate tied in an extravagant bow. His heavy-lidded eyes and thin lower lip make him look more determined than handsome, and a curved scar under his right cheekbone bears witness to an injury of unknown origin. Later portraits show him staring off into the middle distance as if he is seeing great things not apparent to other men, things that he finds unspeakably sad, but this young man looks directly at the viewer in a forthright manner. He has learned a trade, he has a future, and he has fallen in love with a young woman from a respectable family.

Joanna Oliver White was pretty in a way that promised handsomeness rather than beauty in later life. Like most girls in Scotland, which had an advanced approach to education for its time, she had attended school. Intelligent and just as determined as her square jaw indicated, she was a match for an ambitious man.

But their marriage began in difficulty. When Robert and Joanna were married in 1847 after a brief courtship, he was twenty-two; she was nineteen and nine months pregnant. Their first daughter was born eight days after the wedding. Banning by the church was inevitable. Five months later, they appeared at the Session of the Kilmarnock Laigh Kirk and, having confessed their sin, were allowed to rejoin the congregation and have their daughter baptized and named for her paternal grandmother, Elizabeth Hamilton Dunsmuir.

Robert Dunsmuir

Punishment and forgiveness – it is a tradition that has stood the test of time. But young people are often impatient with tradition. They look for ways to change how things are done in their world or they look for ways to escape. In the same year their church welcomed Robert and Joanna back into the fold came news of a way for young Scottish miners impatient with the old traditions to change the course of their lives.

∞

The Hudson's Bay Company (HBC) was a British fur trading business founded in 1670 and in possession of a huge tract of land in British North America. By 1848 that tract of land extended from Hudson's Bay to the Pacific Ocean. HBC forts along the Pacific coast imported supplies and exported furs on sailing ships powered by the wind. So when First Nations people told the HBC in 1835 that there was coal lying on the beaches of Beaver Harbour near the northern tip of Vancouver Island, no one was very interested. Aside from the small amount of coal used by the company blacksmith for his forge, the HBC had as little use for the black rock as did the native people, who used it only occasionally in jewelry.

But a decade later, there were markets aplenty for coal on the west coast. Steam-powered ships were taking over the fighting of wars and the transporting of people and goods. The Royal Navy, then at the peak of its power worldwide, was in the process of converting its fleet to steam power. The navy would require coaling stations all over the world and especially in the remote Pacific Northwest.

At first the HBC employed First Nations people to gather the coal that lay on the surface. But it soon became apparent that in order to supply the markets, the company would have to dig underground to obtain enough coal. For this the HBC needed experienced miners. The company had a longstanding policy of hiring Scots to staff their fur trading posts, for they found them to be hardy people, able to endure hunger, obedient, and prepared to eat poorer food and accept poorer wages than were Englishmen. Not knowing whether these qualities also made for good coal miners, the company ordered mine manager David Landale to find Scottish miners willing to go to the wilds of Vancouver Island, a place most people in Britain had never heard of.

Since miners from Ayrshire had a reputation for producing more coal in a day than miners from any other part of Scotland, Landale hired Ayshireman John Muir as oversman and Muir's four sons and two nephews as miners. They sailed with their families for the HBC's Fort Rupert on Vancouver Island in November 1848.

The Muirs' departure had no immediate effect on the Dunsmuirs. Robert and Joanna, who now called herself Joan, were settling into married life and awaiting the birth of their second child. But when John Muir's sons led the first strike on Vancouver Island in April of 1850, the fates of the two families were joined.

The Muirs were unhappy with the expectations of the HBC. They had been hired to dig coal and yet were expected to sink shafts, something they did not know how to do. Their contracts agreed to pay them on the basis of how much coal they dug, but since there were no pits, they could dig no coal and were therefore earning no money. The company decreed that they would do labouring jobs instead. But these proud Scottish coal miners were unhappy doing manual labour, and so they went on strike.

As the work stoppage dragged on into the summer, the striking miners found themselves in an awkward situation. They wanted nothing more to do with the HBC, but they had no means of leaving Fort Rupert, isolated as it was at the north end of Vancouver Island. Then the sailing vessel *England* entered the harbour to take on coal. The ship rested at anchor while First Nations women filled its bunkers with coal brought in baskets by canoe, and the *England*'s crew filled the miners' heads with tales of the big gold rush in California. When the *England* sailed, she carried all the HBC coal miners and their families, leaving behind only oversman John Muir and his wife and youngest son.

Muir wanted out of his contract. James Douglas, the HBC Chief Factor in Fort Victoria, wrote head office in London for replacements. In the days before railroads and telegraph lines spanned North America, the only way to communicate between Vancouver Island and Great Britain was through letters carried on sailing ships which took four to five months to make the journey. Douglas's letter left Fort Victoria in July of 1850; it arrived in David Landale's Edinburgh office in November.

Despite the company's trouble with the Muirs, Landale was still convinced that Scottish miners were the best choice. He said they were a better class of people and accustomed to doing a longer day's work. And conditions for hiring were good: the Scottish coal trade was in a slump, wages had been cut, and miners were striking in Scotland that summer too. There were sure to be men interested in going to the New World.

In order to reach the HBC ship in time, the new miners had to leave Scotland for London by December 9th. Landale quickly found four miners who would go, but he didn't find an oversman until November 27th, when he recruited Boyd Gilmour. The thirty-five-year-old coal master was an ideal recruit: young, skilled, and married with a small family. The presence of a man's family, the company thought, would make him less likely to desert, although it hadn't worked with the Muirs.

The Gilmour party had just twelve days to settle their affairs in Ayrshire and pack the supplies they would need for a half-year-long voyage followed by life in a primitive fort for the duration of their three-year contracts. Each man received ten pounds (£10) advance money to expedite the process. Then, just six days before departure, Gilmour wrote to Landale in great distress. Someone in the district had received a letter from Vancouver Island that contained alarming news.

"Muirs party are all left the Island and two seamen shot," Gilmour wrote to Landale in words sometimes misspelled, "by orders of a Mr. Blackendish who has since been hanged. The present party are very much dishartened – they think it would be much better to ly in jail in Scotland than ly in irons in Vancouver."

Landale called it a "monstrous absurdity." He denied knowledge of any such events and said that on the contrary "things are going on most harmoniously with all parties." He demanded that Gilmour tell him the names of his informants; he challenged the men to ignore the "silly stories" or hand back their £10 to the town clerk and be done with the venture.

Having sent this reply to Gilmour, he wrote to London enclosing Gilmour's letter and begging the London office to send him a letter denying the truth of the story so he could show it to the miners and "try to dispatch them still."

Gilmour replied to Landale the next day. By this time he had seen the letter from Vancouver Island with his own eyes. Demonstrating a flair for the dramatic that he would use again on Vancouver Island, he told Landale he was "fully determined to go through the undertaking or die in the attempt however disheartening it may be." But two of his recruits returned their £10 and withdrew. With only five days to go, Gilmour had to find replacements. It is some measure of the conditions in Scotland at that time that he was able to find not two but three men who would leave the place where their families had lived for generations and sail for an unknown future in the New World on such short notice.

One of the three men was Robert Dunsmuir. Although Gilmour was his uncle, Robert had not joined the expedition at first. Now, however, despite the rumours of desertion, murder, and hanging on Vancouver Island, he was prepared to go, and so, apparently, was Joan, by now pregnant with their third child. There was so little time before departure that the three new men would have to be given time in London to buy supplies before they boarded the sailing vessel *Pekin*.

The suddenness of the Dunsmuirs' departure for the New World has given rise to the legend that Robert promised Joan a castle if she would come with him. But as they headed for London accompanied by Mr. Landale, who seemed to be going out of his way to assure them that all was well, the only dwelling place the young mother could count on was a tiny cabin in a wilderness fort. To be embarking on a six-month-long voyage halfway round the world to live in such a place was bad enough. It now appeared that the place where they would be living was a lawless and dangerous one.

Young Robert Dunsmuir leaves Scotland for Vancouver Island despite the rumours of murder and desertion at his destination.

2

The Wilds of Vancouver Island

The *Pekin* lets loose her lines at the docks in London. As she traverses the Atlantic Ocean to the tip of South America, rounds Cape Horn to the Pacific Ocean, and sails up the west coasts of South and North America, wind batters her for days on end, then vanishes to leave her becalmed and at the mercy of monotonous swells. Salt water drenches the passengers and crew; fresh water is rationed. When there is enough food, there is not enough variety; when supplies run low, the passengers go hungry and some develop scurvy.

The *Pekin*'s destination is Fort Vancouver, which lies in Oregon Territory inland up the Columbia River. The 148-kilometre stretch of river between the fort and the Pacific Ocean is notorious for the sandbars that lurk and shift beneath its surface. Like so many other ships bound for Fort Vancouver, the *Pekin* runs aground in the mouth of the Columbia. The captain, who should have been prepared for such a mishap, loses his temper instead. Lured by the promise of the easy life that California gold will buy, he and the entire ship's crew, plus one of the miners and the blacksmith, desert the ship.

It is June 18, 1851, and the remaining passengers are stranded. Jean Gilmour and Joan Dunsmuir are both pregnant and their babies are due any day. Then, Peter Skene Ogden, the HBC Chief Factor, comes to the rescue, assisted by the local First Nations people, who unload the ship. Relieved of her cargo, the *Pekin* floats free of the sandbar and negotiates her way gingerly up the Columbia toward the fort. Enroute Jean Gilmour gives birth to her sixth child, a son whom she calls Allan Columbia; on June 29th the *Pekin* ties up on the northern shore of the river at Fort Vancouver.

Before the Oregon Treaty of 1846 set the border between the United States and British North America at the forty-ninth parallel, 375 kilometres to the north, Fort Vancouver was the headquarters of all HBC activity west of the Rocky Mountains. But now, although the fort continues in the service of the company, the new headquarters is within British North America at Fort Victoria on the southern tip of Vancouver Island.

Although it has grown shabby, Fort Vancouver's stockade is large and strongly built, and the fields outside the walls grow abundant crops in the rich soil of the river delta. Ogden and the staff of the fort welcome the Gilmour party warmly. Here, amid the security and frontier abundance of an isolated and somewhat dilapidated American fort, Joan Dunsmuir gives birth to her third child and first son, James, on July 8, 1851.

At the same time, far to the north, Her Majesty's Ship (HMS) *Daphne* drops anchor near a Nuwitti village about forty kilometres from Fort Rupert.[1] Her skipper, Captain Fanshawe, under the supervision of Rear-Admiral Fairfax Moresby, who is also aboard, orders sixty Royal Navy marines over the side into smaller boats. Their mission is to proceed to the village and seize the murderers of the three sailors who escaped from the vessel *England* the year before and whose murders caused such anxiety when Boyd Gilmour was recruiting Scottish miners for his crew.

The three escaped sailors had taken refuge on an island near Fort Rupert. One day, they saw a party of friendly Nuwitti approaching by canoe, but they misunderstood the visitors' motives and threw rocks and threatened them with an axe. Friendly no more, the Nuwitti shot and stabbed the Englishmen, then stripped them of their clothing and tried to hide their bodies by sinking at least one in the ocean and stuffing another one upright into a hollow tree.

Now, a year later, the Royal Navy has come to punish the Nuwitti murderers. In the raid on the village, the marines destroy twenty canoes, set fire to the houses, kill six Nuwitti including the chief, and leave five of their own dead on the beach. They return to the *Daphne* empty-handed.

1. The Nuwitti nation is a member of the Kwakwaka'wakw ethnic group, formerly known as Kwakiutl or Kwagiulth, whose traditional territories cover northeastern Vancouver Island and the adjacent mainland.

Having punished the village if not the murderers, the *Daphne* leaves Fort Rupert on another mission. Then the Nuwitti, according to their custom, turn the murderers over to the HBC in return for a promise of thirty blankets. When the Gilmour party arrives four weeks later, they know nothing of the deaths at the Nuwitti village, but the murder of the sailors the previous year, which clouded their departure from Scotland, is still vivid in their minds.

∞

At Fort Vancouver, the Gilmour party with its two new babies boarded the HBC's ship, *Mary Dare*, for the voyage up the west side of the Olympic Peninsula and Vancouver Island and around the Island's northern tip to Beaver Harbour.

A scattering of steep-sided islands guarded the mouth of the harbour, which was otherwise open to the north. Passengers standing at the *Mary Dare*'s starboard rail and scanning the horseshoe-shaped bay saw how the grey sea met the tree-covered land in a continuous arc of wilderness. It was not until their gaze reached the south shore that they saw puny evidence of civilization. There, dwarfed by a backdrop of evergreen forest, was a roughhewn stockade with a bastion at each front corner. Butting up against its picketed walls were several long, shed-like buldings.

Because the harbour bottom sloped gradually away from shore, larger vessels had to anchor some distance out. The *Mary Dare* fired the two-gun signal of HBC ships, set her anchor, and lowered her tender to transport the passengers and cargo ashore in shifts. When Joan and Robert Dunsmuir stood at last on the shore with their four- and two-year-old daughters and their infant son, they had been enroute from Scotland for almost eight months.

∞

The first thing the newcomers heard was of the events of the past year. What Boyd Gilmour had feared and David Landale had called a "monstrous absurdity" was even worse than the rumours had said: three men, not two, had been murdered. No one had been hung, it was true, there being no such person as Mr. Blackendish, but drunkenness and insubordination had destroyed discipline at the fort for months on end. Then, just weeks before, the Royal Navy had exacted vengeance on the Nuwitti village.

But now it was early August, 1851, and only the arrival of the Gilmour party disturbed the midsummer calm. The inhabitants relished the August sun as it offered brief respite from the normal rain and fog. In the absence of the chief trader, two young and inexperienced men – George Blenkinsop and Charles Beardmore – were in charge of the fort as they had been the year before when the fort was anything but calm.[1] Life was now peaceful, though the murders of the past year still haunted the inhabitants and the thousands of Kwakiutl whose longhouses hugged the outer walls of the fort.

1. The name of the mythical Mr. Blackendish was probably a blend of "Beardmore" and "Blenkinsop."

21

The Kwakiutl nation was part of the same ethnic group as the Nuwitti and so shared the same territory, language, and culture. Lured by the blankets and tobacco they could earn by digging and loading surface coal and by supplying the fort with fresh meat and fish, the Kwakiutl had lived around Fort Rupert since 1849. It was beginning to dawn on the HBC men that though the Kwakiutl were warriors and dealt in the capture and sale of slave labour, they had their own system of justice and were not interested in harming the inhabitants of the fort.

The newly arrived coal mining families knew nothing of this. To them, the "Red Indians" were a dirty lot, their faces dark and painted with streaks of vermillion, their naked bodies covered only with blankets. Their village bristled with the spears, knives, and flintlock muskets they used in their slave-taking expeditions.

At the end of these expeditions, the harbour filled with returning war canoes and the beach with severed heads impaled on wooden stakes. Old hands reassured the newcomers that this was merely a display of honour. Two years before, Kwakiutl warriors had offered John Muir's wife, Anne, her choice of any two heads because she was the first white woman they had ever seen. They showed Agnes Hunter, a more recent arrival, similar homage when they threw a bloody head at her feet as she walked along the beach.

It was honour and not terror that the Kwakiutl wished to convey. As John Helmcken, a young doctor who had been at the fort during the recent troubles, said of them, "among these people we walked and

Laird of the Mines

roamed and certainly, after having become accustomed to them felt less fear of molestation than I had often experienced when traversing the slums of London."

Friendly though they were, the Kwakiutl had been angry when the Muirs started to dig for coal three years before. The Kwakiutl men had been digging surface coal for ten years at Suquash, ten kilometres south of the fort, using trade axes, hammers, and crowbars to remove the trees and earth that lay over the coal close to the surface. Their women had paddled coal-laden canoes out to passing ships in exchange for trade goods. When Muir and his men began to dig, the Kwakiutl men, thinking their source of income endangered, had surrounded the workings and threatened to kill the Muirs.

But the Muirs, having found so little coal, had proved to be no threat. When the entire family, except John and Anne Muir and their youngest son, Michael, fled the fort for California, eight hundred Kwakiutl had continued to dig the surface coal at Suquash. Muir, on the other hand, had been unable to locate a viable underground coal seam and had left Fort Rupert months before Gilmour arrived.

In the spring of 1851, while the fort waited for the new miners, the HMS *Tory* arrived with twenty-five labourers, a steam engine, and rods for drilling. The new men set to work clearing Muir's pit in preparation for Gilmour's arrival. If Gilmour succeeded where Muir had failed, it would again threaten the Kwakiutl workforce. But first, Boyd Gilmour had to find a workable body of coal.

23

HBC employees and their families lived inside the four-sided stockade in rudimentary houses whose only advantage was that they were an improvement over the cramped quarters aboard ship. Clay-chinked log walls supported roofs with a hole in the middle to let smoke out and fresh air in. Since it often rained, the hole also let water in, water which drizzled onto the bare earth floors, whose covering of fragmented clam shells did little to prevent the formation of mud. One bake oven in the centre of the fort's inner courtyard served all the houses.

Joan Dunsmuir and Jean Gilmour were accustomed to proper houses with chimneys, ovens, and floors. Their homes in Scotland had had kitchens, their town greengrocers and butchers. Now, if their families were to eat, the Scottish women had to skin and butcher deer, gut fish, pluck wild fowl, and tend vegetable gardens. If their laundry was to get clean, they had to make soap. In order to learn these skills, they had to befriend the native women from the tribes farther north who, having married HBC employees, lived inside the fort too.

While the women adjusted to bloody deer carcasses and muddy children, the men had to find coal. Boyd Gilmour, in particular, had to justify the company's faith in him. Having dismissed Muir's shaft near the fort as useless, Gilmour and his small crew of twenty miners and labourers moved the new drill rods to Suquash.

Among the English labourers now working for Gilmour was a man who had been at the fort for over a year. Edward Walker was a jack-of-all-trades, able to do whatever was asked of him. California's gold fields held no fascination for him. He knew that employers valued loyalty and hard work. Robert Dunsmuir knew this too. Being the same age and similar in outlook, the two men became friends.

But having a congenial working companion did not make the work any less frustrating. By the following spring the Gilmour crew had been no more successful at finding a workable coal seam than the Muirs had been. Miners and labourers had been deserting all winter. Chief Factor Douglas, who had described Gilmour as "not being very sanguine but full of ardour," was beginning to wonder if Gilmour had any idea what he was doing.

Like John Muir, Boyd Gilmour was not qualified for the job he was trying to do – find coal and develop a mine. In a determined effort to justify himself, Gilmour decreed that he and his rapidly shrinking crew would drill elsewhere: behind the fort, farther inland southwest of the fort, twenty-five kilometres along the coast. They found quicksand, they found narrow fifteen-centimetre coal seams, but mostly they found whinstone, a hard blue rock which was so difficult to penetrate that a full day's drilling only yielded them a hole twenty centimetres deep.

In Gilmour's opinion, there was no coal underneath the whin, and he saw no point drilling any deeper. John Muir, who was now advising Chief Factor Douglas in Victoria, disagreed with Gilmour. In his opinion, the whin was just a dyke intersecting the coal bed, and there was a good chance that coal lay beneath it. But since neither man had found any useable coal on either side of the whin, Douglas didn't know who to believe.

"Gilmour rather dispirited over his want of success," Douglas wrote in his sloped handwriting to the head office in London. "The miners have of late shown frequent signs of discontent." Their lack of success had been made worse by a shortage of money, the eventual arrival of which "gave great satisfaction and removed every trace of despondency."

Through all the frustration and desertions, Dunsmuir stayed loyal to his contract and to his uncle. He watched and learned from the older man, but drew his own conclusions about the nature of the local rock formations and how to find coal. Coal usually shows itself where the underground seam, lying at an angle, breaks through the surface of the earth in an outcrop. Sometimes it reveals itself in the roots of a fallen tree; sometimes lying loose on a beach. Drilling into the earth at the site could confirm the existence of a workable seam of coal. Dunsmuir was well-schooled in the art and science of mining coal, but still had much to learn about finding it, especially on Vancouver Island. Though he had formed opinions, the time for him to act on them had not yet come.

Boyd Gilmour could not afford to be so patient. By the summer of 1852 he had concluded that the only coal around Fort Rupert was surface coal. The Kwakiutl, not wishing to work at depths of more than three metres, refused to continue working. Then word from the south offered a way out for Gilmour and the HBC.

Laird of the Mines

Just that summer, two and a half years after a man of the Sne ney mux nation had reported the existence of coal on the beach near his home on the shores of Winthuysen Inlet, James Douglas had travelled north one hundred kilometres from Fort Victoria to inspect. This time there was lots of coal – a naval vessel had already taken on 480 barrels of the black treasure. The HBC had established a new fort on the shores of the sheltered and welcoming harbour. The fort's name was Nanaimo, the HBC version of Sne ney mux. If there was coal at Nanaimo, then that was where Boyd Gilmour wanted to be.

Jean Gilmour had had enough. A year in the wilderness was all she could tolerate. Her husband might well be going to Nanaimo, but she was going to go to Victoria to live in comfort until his contract expired and they could return to Scotland. Joan Dunsmuir, newly pregnant with her fourth child, would stay with Robert.

Winter rain and fog had settled on the fort as the time drew near for Gilmour and his crew to leave. That was when Edward Walker broke his leg in a mining accident. The injury was so severe that he had to stay in bed. But Gilmour was anxious to be in Nanaimo, word having arrived that his rival, John Muir, whom he had never even met, had rejoined the company and had been there since September. The record shows that Gilmour left in December, but it does not show when Dunsmuir left.

It would be natural to assume that Dunsmuir left at the same time as his uncle did, but some sources say he didn't go until February 1853 when his son, James, was nineteen months old. Others say he waited until April when his friend Edward Walker was able to travel.

Robert Dunsmuir's name is seldom mentioned in the HBC records; their pages are filled mostly with troublemakers and crises. But since Walker was bedridden and Dunsmuir was his friend, it is reasonable to suppose that Robert Dunsmuir stayed behind in Fort Rupert until Walker was able to travel. The friendship that lasted between the two men all their lives could well have been cemented with just such an act.

The decision to stay behind, if taken, would have served Dunsmuir well. While he was still in Fort Rupert, the miners and labourers working under Boyd Gilmour in Nanaimo went on strike. James Douglas, who visited Nanaimo shortly after the strike began, was impatient. "[Gilmour] was at variance with the miners under his charge ... They brought forward a long string of petty complaints, which I disposed of in a very summary manner, and soon brought them to their senses."

Gilmour seemed to be having as much trouble with his employees in Nanaimo as he had had in Fort Rupert. His nephew, on the other hand, seemed to know the value of keeping his head down and causing no trouble. A man who causes no trouble is valuable to an employer.

Without Joan Dunsmuir's strength and determination, her husband would never have been as successful or as ruthless.

3

So Useful a Man

On the waterfront in Nanaimo in a rough stone cottage, Captain Charles Stuart sits at his HBC office desk writing. The post journal lies open before him waiting for its daily ration of babies born, coal mined, deserters punished, and deaths investigated. It is October 1855 and he has been the officer in charge at Nanaimo for two months.

Stuart is thirty-eight years old, a HBC mariner before taking this job on land. He has as little experience in running a settlement as his employer has in mining coal. The former ship's captain has to supervise the building of houses, wharves, and mine scaffolds; oversee the mining, loading, and selling of coal; arbitrate disputes; punish criminals; conduct burial services; and, as the representative of the only property owner, be the only Nanaimo voter in colony elections.

If it is possible for one man to do all these things, Stuart is such a man. He is energetic, suave, warm, and open-handed. But he has been sorely tried in his short time as officer in charge as he struggles to run a coal mining business in the tiny village so far from the civilized world.

Not that there is any shortage of coal in Nanaimo: there are mines scattered all over town. But mining coal requires manpower – miners to dig coal and labourers to haul it. The labourers available to Captain Stuart are either employees of the HBC or men and women from the Sne ney mux nation. Although the Indian women are dedicated workers, their men are unreliable by HBC standards, as likely to go hunting as to show up for work underground. And the Sne ney mux have other concerns. Often and without warning, Haida or Kwakiutl war parties storm into the harbour intent on taking Sne ney mux captives or severing Sne ney mux heads from their bodies.[1]

The company labourers are even less dependable than the local First Nations men. Put an Iroquois or a Canadian from Lower Canada or a Kanaka from the Sandwich Islands or an Englishman, for that matter, in a mine and they soon think themselves experts. Chastise them for laziness or neglect of their duties and they are likely to desert for the mines at Bellingham Bay or the newly discovered gold fields on the Fraser River on the mainland.

1. The Sne ney mux nation inhabits the area around Nanaimo and Gabriola Island and is part of the Hul'qumi'num ethnic group.

Laird of the Mines

Much worse than the labourers, however, are the miners. Seldom in the 180-year history of the HBC has there been a more bothersome lot of employees. The miners and their oversmen have been causing trouble in Nanaimo ever since the first mine opened in September, 1852. Muir and Gilmour competed and complained until the company was glad to see the end of their contracts. Only two of the twelve Scottish miners recruited by the HBC have stayed in Nanaimo after their contracts expired: Muir's nephew, John McGregor, and Gilmour's nephew, Robert Dunsmuir. Both men have proven valuable to the company it is true, but even they have caused their share of trouble.

The company, having no wish to hire any more troublesome Scottish miners, has hired Englishmen from the Black Country in Staffordshire. But they are no better. Ever since they landed in Nanaimo in November, 1854, they have complained, refused to work, gone on strike, and deserted over and over again.

∞

There were such great hopes on the day they arrived. The people of Nanaimo – one hundred and fifty-one of them and "a goodly number of Indians" – stood under a cloudy November sky in 1854 to greet the twenty-three Staffordshire miners and their families. Behind the Nanaimoites was a solitary row of houses: cedar shingled, sooty, squarehewn log cabins in dire need of their regular coat of whitewash. To the right of the houses was a single bastion showing the only flash of colour in the otherwise dingy scene: a Red Ensign with "HBC" on the fly furling and unfurling in the wind.

Robert and Joan Dunsmuir and their children shared one of the cabins – two rooms warmed unevenly by a fireplace. Until recently, they had also shared it with Boyd Gilmour. It was necessary, given the shortage of living space, but it had satisfied no one. Gilmour had complained officially to Chief Factor James Douglas, by now also Governor of the new British colony of Vancouver Island. Douglas had written to Joseph McKay, clerk in charge at Nanaimo, that Gilmour "appears much dissatisfied with his lodgings, and his treatment generally, as he evidently considers himself slighted." Four months later Douglas was still trying to pacify Gilmour by trying to get him a small amount of wine. Since Douglas had imposed heavy duties on alcohol in the colony in order to curb drunkenness, this was a very special favour.

The favour came after Gilmour had finally found some coal. But that success did not change his decision to return to Scotland when his contract ran out a year later. He packed his belongings, moved out of the cramped cabin, and left for Victoria to join his wife and children. There they waited until Jean gave birth to their seventh child before they sailed for Britain on the return voyage of the *Princess Royal*, the ship that had brought the Staffordshire miners from England to Victoria.

According to the Staffordshire miners' contract, they were to build their own houses with materials supplied by the company. But these men had no idea how to build a house, especially one of squarehewn logs.

Even if they had known what to do, they needed shelter immediately. Since Nanaimo was already suffering a housing shortage, families crammed themselves into whatever space they could find.

And the space was usually in need of repair: wind howled through the cracks between the logs where the chinking had fallen out. Benches and bunkbeds, spare and plain, were all the furniture they had, and coarse drugget mats only partly covered the bare earth floors. The dim light of small fish oil lamps and the flickering warmth of the fireplaces were puny against the dark and damp of short winter days and long winter evenings.

Having suffered a voyage more gruelling than most, the Staffordshire miners arrived unhappy and stayed that way. Within two months, an American coal company had lured six of them with "flattering tales" and free whisky to leave Nanaimo and come to Bellingham Bay in Washington Territory. Thus began a cycle of desertions and work stoppages which lasted for at least four years.

The combination of a shortage of labour and an employer inexperienced in mining could only lead to trouble. The promise of better conditions and higher wages at Bellingham Bay easily lured men away. And the men who stayed behind were just as much trouble. Half of them refused to go to work and the other half only pretended to dig coal. Their behaviour drove George Robinson, the new mine manager, to distraction. As he walked through the tunnels and crosscuts of each mine, he challenged them in vain to dig a decent day's tonnage.

"One part of them simply laugh at you for doing so," he wrote to Douglas in his quarterly report, "and the others will make use of the most offensive and insulting language imaginable."

Having found conditions in Bellingham Bay to be even worse than in Nanaimo, the deserters eventually straggled back. Then the complaining began anew. In September, 1855, the discovery of gold on the Pend d'Orielle River in Washington Territory gave eight miners a reason to lay down their tools and refuse to work. The next day, one of their number, John Meakin, having consumed far too much whisky, loaded his shotgun and aimed it at his wife.

As the domestic drama moved into the street, onlookers sent for Captain Stuart, who had been in charge at Nanaimo for only a few weeks. They demanded he arrest Meakin and lock him up in the basement of the Bastion, but Meakin's fellow Englishmen thwarted Stuart on every hand. Stuart attempted to handcuff the young man. Meakin's friends dragged him out of Stuart's reach and demanded that he fight Meakin one on one. Determined to maintain his own dignity and that of the company, Stuart ignored them. "The conduct of *all* of the Englishmen and one or two of the Scotch miners was disgraceful in the extreme," he confided to his journal in his pale, neat handwriting later that night.

The only men, it appeared, who were keeping their heads and going about their work were Edward Walker and Robert Dunsmuir. Walker seemed to be able to do anything the company asked him to do: build mine scaffolds, fell logs, dig coal. He bought a boat and filled it with refuse coal to take to Victoria for sale; he returned with a load of flour and potatoes; he carried dispatches between the two forts; he installed beacons in the harbour.

His friend, Robert Dunsmuir, had been making himself useful in other ways. Dunsmuir had come to the notice of Chief Factor Douglas, who saw him as a steady, intelligent man. In August, 1853, just months after Dunsmuir's arrival in Nanaimo, Douglas had put the young man to work proving the coal his uncle had discovered at the head of Commercial Inlet.

The coal lay at the point where the inlet narrowed and curved to the north to become a scrawny finger of tidal water which divided a fat peninsula of land off Vancouver Island. It was on this peninsula that Nanaimo stood. Dunsmuir's job had been to measure the thickness of the Douglas seam and drill through the conglomerate rock that lay below it to determine whether there was a second seam.

The Douglas seam was two metres thick at the head of the inlet. From August to December, 1853, Dunsmuir had drilled through the seam and then had continued to drill painstakingly through twenty-five metres of conglomerate rock, sandstone, and shale until one day, four days before Christmas, he had hit the Newcastle seam.

There was so much coal and so few men to work it. No wonder the company took the Staffordshire men back time after time. It made Robert Dunsmuir an even more valuable employee. Here was a man who not only could find coal, but who avoided trouble at all costs.

Robert Dunsmuir

⚭

Charles Stuart continued to thumb through his journal entries for the three months since his arrival. October 1, 1855: "Mrs. Robinson delivered of a living male child." October 6th: "The miners returned from Victoria by canoe having been reprimanded by the Governor and engaged upon less favourable terms." October 12th: "Dunsmuir commenced working on his own account."

Robert Dunsmuir and Edward Walker, having completed their contracts the summer before, had now been granted the first free miner's licences ever bestowed by the company. No longer would either man have to bow to the company's will. And while each would be dependent on the company for work, there being no other source of income in Nanaimo, Walker and Dunsmuir would succeed or fail on their own merits.

Walker's new status bothered no one. The work he continued to do – screening coal, delivering lumber, fitting guns on the Bastion – threatened nobody. But Dunsmuir's new status seemed to anger the Staffordshire miners. While they continued to argue about their pay and working conditions, he could set his own. It maddened them that he had refused to go along when the miners deserted. Was he just a sensible man who knew how to avoid trouble? Or did he have a secret plan? Had he gambled that he would get a free miner's licence if he refused to join the troublemakers?

Whatever Dunsmuir's reason for remaining loyal, the company was grateful. The chief factor granted him a long-term contract with the Nanaimo Coal Company saying, "[I]t is the Hudson's Bay Company's interest to openly encourage such enterprises whereby a steady miner might by honest industry improve his condition."

Later in the month the reasons for grumbling seemed unimportant when word came that Kwakiutl war canoes were in the Strait of Georgia heading south. Two islands, Newcastle and Protection, form the outer edge of the harbour and shelter Nanaimo from the storm-bearing winds. But since the islands also block the view of the Strait, no one was certain whether the Kwakiutl canoes were headed for Nanaimo until they swept through the gap between Newcastle and Protection. Then more canoes appeared – their prows thrusting upward, their crews digging short-handled paddles into the water on one side then the other – advancing down the channel that connected the harbour to Departure Bay.

"Women and children to the Bastion," rang out the call. Someone fired a few warning shots from the cannons, but the Kwakiutl weren't interested in the people rushing to the shelter of the small fortress. As was often the case in disputes between the various First Nations, they sought revenge against each other. This time the Sne ney mux had murdered three Kwakiutl. One, two, three days passed, three days of consultations while the white women and children huddled in the Bastion. Finally, the two sides agreed on a solution: a Nanaimo chief would be executed in retribution for the murders. The execution took place and the Kwakiutl climbed back into their canoes satisfied that justice had been done.

Robert Dunsmuir

As far as George Robinson was concerned, it was just an annoying interruption in the work of the mines. "Had to suspend workings in consequence of the excitement produced amongst the Indians by the assassination of Won Won Chin (a Sne ney mux chief) by some of the northern Indians," he reported to Douglas.

It was the only time the Bastion served its original purpose – protection from Indians – and it turned out to be merely a minor inconvenience for everyone except the unfortunate chief. But the petty annoyances continued as the two cultures tried to adapt to each other. The Sne ney mux were used to living communally in multi-family longhouses; they saw nothing wrong with walking into the miners' houses unannounced. The Sne ney mux were used to sharing goods and giving gifts; they saw nothing wrong with taking the miners' belongings if something caught their fancy. And a dark-skinned people saw nothing wrong with staring in fascination at the pale-skinned babies of the European settlers.

Joan Dunsmuir came into her cabin to find a knife-bearing Indian man standing over the cot of her newborn son, Alexander. The man turned and looked at the appalled mother, grunted and walked out. This event complemented a story about Alex's older brother, James, which became part of Dunsmuir family lore. Succeeding generations of Dunsmuirs told how the Kwakiutl at Fort Rupert had kidnapped the small blond baby boy when the family lived there. The story of how his frantic parents found him after a few hours and how the Kwakiutl then offered to buy him for sea-otter skins piled "to the height of a man" never appeared, however, in the official records.

Laird of the Mines

Having her children kidnapped by Indians may not have been what Joan Dunsmuir expected when she agreed to come to the New World. But unlike Robert's Aunt Jean Gilmour, Joan was prepared to make the best of things. Her family say she took in washing when she first came to Nanaimo. Whether or not such hard labour was necessary given her husband's success in finding coal, the family believes that she was the stronger of the two Dunsmuirs, and her behaviour in later years does not contradict this.

Robert Dunsmuir's new enterprise was indeed successful. In February of 1856, Manager Robinson informed Chief Factor Douglas that Dunsmuir's Level Free mine – so named because gravity drained the water from its upwardly inclined slope – contained an extensive amount of moderate quality coal. Dunsmuir and the men working for him had dug so much coal that he had been able to fill the HBC ship *Otter* entirely with coal from his mine. And because Dunsmuir had taken special pains to clean his coal by screening the shale from it, the markets in San Francisco said, "Send us Dunsmuir coal."

Robinson watched Dunsmuir's success with mixed feelings. On the one hand, the Scottish miner had certainly dug plenty of clean coal; on the other, he was one of the few people not under Robinson's control.

Robinson was more qualified than his predecessors, Muir or Gilmour, to run a coal mine, but he was no better at employee relations. A man with a quick

Robert Dunsmuir

temper, Robinson also had good reasons to be out of sorts. Since arriving in Nanaimo, both his wife and baby son had died. Added to the strain of this tragedy was the continual round of desertions and complaints that plagued him as he tried to make the Nanaimo Coal Company into an efficient producer of coal. The strain showed in an altercation he had with John McGregor, which ended in him hitting the rangy oversman over the head with a blacksmith's hammer.

Contrite as Robinson may have been when he reported his lapse to Stuart, his treatment of Robert Dunsmuir showed that he was still capable of foul behaviour. Dunsmuir's mine was the only one that continued to work during labour disputes. Dunsmuir's coal was what the markets wanted. Dunsmuir's discovery of an extensive new fold of coal forced Robinson to order every available miner, some of them very reluctant to work for an independent operator, into Dunsmuir's mine.

Spitefully, Robinson advised Douglas that the company was paying too much for Dunsmuir's coal; Douglas cut the price he paid to Dunsmuir by three shillings a ton. Dunsmuir charged that while he was away, Robinson loaded a ship with his coal and underestimated the tonnage to cheat him of his rightful profit. Douglas tried to be impartial. When Dunsmuir spoke rudely to Robinson, Douglas sympathized with the manager, but worried about losing the services of "so useful a man."

In the contest between the two men, however, Robinson had the upper hand. He gave orders that a company mine be opened in the same area as

Laird of the Mines

Dunsmuir's at the head of Commercial Inlet. Dunsmuir had predicted that his seam would last four more years. With the new mine eating away at the same coal, it would never last that long.

Later a company official would acknowledge that Robinson had been motivated by his hostile feelings toward Dunsmuir. "There is no doubt that it was a great error to open the slope, Dunsmuir being capable of working the whole."

Whether they resented his capabilities or his decision to avoid trouble, the workers assigned to his level free mine grumbled continually. But Dunsmuir needed them if he was to work his coal. Until the discovery of gold in 1858 on the Fraser River brought thousands of gold miners into the colony and the unsuccessful among them provided a large pool of available labour, Dunsmuir put up with the grumbling.

In the year of the gold rush, Robert Dunsmuir built a house, and his nemesis, George Robinson, bought a camera. In a photograph taken by Robinson, the new Dunsmuir home looks down on the rest of the village from the other side of the Ravine, as Nanaimoites called the tidal inlet. With a front porch and a peaked roof that swept down at the back like a sou'wester hat to cover a shed-like kitchen, it gave the added room needed to hold the Dunsmuirs' growing family, which now included another daughter, Marion, born two years before.

It looked as if the Dunsmuirs planned to stay in Nanaimo. But changes were in the air. The company fired Captain Charles Stuart for chronic drunkenness. George Robinson's contract expired in 1859, and he

Robert Dunsmuir

left the Island temporarily for the more civilized delights of England's County Sussex. Robinson's lack of popularity in Nanaimo was apparent when a vague proposal circulated that his departure be celebrated by hanging him in effigy.

∽

When Dunsmuir's Level Free mine ran out of coal, he ordered his miners to "draw the pillars." Working back from the face,[1] the men set blasting powder charges in the pillars of coal that had been left to hold up the roof of the mine. As the roof gradually fell, the rock surrounding the mine sometimes cracked.

One day when Dunsmuir's men were taking a break from their work, they lit a small fire at the entrance of the mine to warm themselves. At the end of the shift they left the fire smouldering. During the night a strong wind rose and fanned the fire, which crept back into the mine, igniting the waste coal and the old timbers.

The next morning the mine was in flames. Workers worked frantically to close up the entrance to deny the fire oxygen. But oxygen was getting in through the cracks in the roof. The coal burned for many years, the heat it produced expanding the cracks and allowing smoke and steam to escape to the surface.

But even as the smoke and steam seeped out of the abandoned mine, a new era began for Robert Dunsmuir. The HBC sold its mines to the Vancouver

---

1. The face is the wall of coal at the end of a tunnel or miner's stall into which miners drill and set powder shots to loosen coal.

Coal Mining and Land Company (VCML), and Robert Dunsmuir, the man chosen by the new company to be superintendent, was the best man possible for the job.

Nanaimo in 1858 reveals the Bastion on the shore of the harbour with the first miners' houses nearby on the left. Dunsmuir's new home is in the foreground.

B.C. Archives 11290 (b)

4

For He's a Jolly Good Fellow

The new deep-water wharves of the VCML are a wonder to behold. From their foothold on a peninsula sticking out into the harbour, the wharves seem to tower above the ships waiting to take on coal. Tall pilings support two platforms thirty metres wide and one hundred fifty metres long where the low-slung longhouses of a Sne ney mux village once hugged the shore.[1]

Just ten years before, native women had paddled canoes loaded with baskets of coal out to ships at anchor. Now the saddle-tank locomotive *Pioneer* pulls twelve wagons, each loaded with four tons of coal, from

---

1. In 1862, as a result of a smallpox scare, the Sne ney mux were persuaded to move to a new village farther south in the harbour.

Robert Dunsmuir

the mine to the wharves several times a day. Attached to the sides of the giant platforms and projecting down toward the water are several chutes, which rise and fall on the tide. The coal pours down the chutes into the hatches of colliers, as coal ships are called, tied up at water level.

The completion of the wharves in 1863 has brought the industrial age to the west coast. Since then, the chuffing of the *Pioneer* and the clatter of the coal as it shoots down into the colliers have become familiar sounds. The *Pioneer* is the first steam engine west of Ontario. People who have never seen one before don't believe that she is stronger than a hundred horses.

After she dumps each load into a waiting ship, the *Pioneer* returns to the Douglas Pit at the head of Commercial Inlet for more coal. The mine is the first one in Nanaimo to have a proper tipple, a tipple so high it seems to rival the ancient firs and cedars growing behind it. A criss-cross of wooden beams supports the hoisting wheel at the top and, farther down, an elevated platform where coal-laden cars roll off a narrow cage that has brought them up from the bottom of the shaft.

Each open car coming off the cage has a copper tally hooked to it just inside at the top, and each tally has a number on it belonging to the miner who dug the coal it carries. Each car moves by gravity along a tramway toward the scales where a weighman records its tonnage beside the miner's name.

It is a noisy and dusty business weighing the cars and tipping them onto a series of screens to sort the coal lumps according to size. The fine coal dust, or dross, stays behind to fuel the engines and the ventilation furnace. The lump coal drops into the wagons hooked up to the locomotive waiting to make another trip to the wharves.

The dust-smeared workers on the wharves, the men and boys working the pithead, the miners and backhands waiting to enter the cage for a descent to the bottom of the twenty-fathom-deep shaft, all have their foremen and pitbosses. But Superintendent Robert Dunsmuir is in charge of them all.

Dunsmuir's main task is to push the main road or tunnel farther and farther into the rich body of coal. Miners scramble aboard the empty cage, the gate swings shut, and the cage plummets to shaft bottom where the huff of steam engines, the muffled thump of blasting, the snort and whinny of horses, the bray of mules, and the rattle and clank of chains and machinery fills the coal-dust-laden air.

An underground locomotive or locie waits with a string of sooty cars to take the miners to the face. Quieter sounds now – wheels clicking over the rails, men coughing and muttering while their fish oil lamps cast a soft light which moves along the walls and timbers of the tunnel as the locie pulls them farther and farther into the mine.

The light comes from Kilmarnock lamps, large-spouted teapots which hook onto the front of the miners' soft peak caps. Wicks made of threads woven together by the miners' wives from castoff cotton clothing burn oil extracted from the livers of cod fish.

When ignited, the oil produces an open flame, which gives good illumination but can cause explosions if it comes in contact with the methane gas or firedamp that bleeds out of newly disturbed coal. In shallow mines like the old HBC pits, fresh air from the outside got into the workings easily and flushed the gas away. In deeper mines like the Douglas Pit, fresh air has to be sucked into the mine with the ventilation system and driven into all the tunnels and crosscuts in order to sweep away the firedamp before it can ignite and explode.

If used properly, the Douglas Pit ventilation system is effective. A fire burns in a furnace at the bottom of a second shaft connected to the mine. The furnace fire, hungry for oxygen, sucks fresh air toward itself down the main shaft and creates a stream of air that can be directed into all the nooks and crannies of the mine where gas may be lurking.

∽

The work was advancing well. More and more ships waited in the harbour to take on Douglas coal for the San Francisco markets. Then, just when his future with the company seemed a foregone conclusion, Dunsmuir received an intriguing offer from his friend, Dr. Alfred Benson.

Benson had developed a reputation for being a bit of a character since his arrival in Nanaimo in 1857 to be the company doctor. He wore a pair of sea-boots with one pant leg tucked in and one flapping out. "Ah," he had said to his colleague John Helmcken, "you laugh, but if you were to remain here a few months you would of necessity become the same."

Laird of the Mines

By 1863, Benson was a private practitioner. His wife of two years having died that spring, he filled his spare time looking for coal outcrops beyond the boundaries of the company lease. When he found coal that fall, he set about looking for investors and found one in the seventh son of the Earl of Harewood, a twenty-eight-year-old lieutenant-commander in the Royal Navy stationed at the Esquimalt naval base outside Victoria.

Like many young British aristocrats, the Honourable Horace Douglas Lascelles had money to invest. The Harewood Coal Company was born in December, 1863, and faced a government-imposed deadline of eighteen months in which to develop the field. What Lascelles and Benson needed was a man who knew what he was doing. Lured from his present post with the promise of a share in the profits, Robert Dunsmuir agreed to join the new company.

∞

John Meakin had reason to be proud. His fellow miners had chosen him to be their spokesman. The man who, when he first arrived on the Island, had threatened his wife with a shotgun in a drunken rage, was now the man who would present the miners' testimonial to Robert Dunsmuir, the very person he refused to work for just eight years before.

A lot had changed in Nanaimo since the HBC sold its land, houses, store, and mines to the VCML in 1862. "The Vancouver Coal Mining and Land Company has animated an enterprising spirit," reported *The*

51

Robert Dunsmuir

British Colonist, a Victoria newspaper that usually looked down its nose at Nanaimo. New houses, new stores, two hotels, two bakeries, and one butcher shop made Nanaimo look more like a prosperous town and less like a struggling fort. The new Institute Hall, while still in need of interior finishing and exterior painting, provided a place where people could gather for important occasions.

And what more important occasion could there be than the Public Testimonial Tea for Robert Dunsmuir on the occasion of his leaving the company? John Meakin stood in front of the assembled miners and their wives and children. He nodded respectfully to Robert and Joan Dunsmuir and their six children, Elizabeth, seventeen; Agnes, fifteen; James, thirteen; Alex, eleven; Marion, nine; Mary Jean, two; and Emily Ellen, a babe in arms. The assembled crowd had eaten well of the food prepared by the miners' wives and now waited for Meakin to speak.

"As a token of the great respect we entertain for our late overseer," Meakin told Mr. Dunsmuir, the miners had collected money and purchased a gold watch and chain with the inscription, "Presented to Robert Dunsmuir by the miners of Nanaimo as a token of respect."

Dunsmuir rose to accept the watch. "When I was amongst you," he said in words tinged with irony, "I little anticipated this kindness, or that I had gained so much of your respect as exhibited towards me this evening, and of which I feel justly proud."

Three cheers went up for Mr. Dunsmuir followed by a round of "For He's a Jolly Good Fellow." Then, while a swarm of local children were allowed in to finish up the remnants of the food, the party-goers began to dance, shaking the unfinished hall until after midnight.

By the time Dunsmuir joined the new company, it had already hired fifteen men to develop the two-metre-thick seam. By July, he had enough coal to take 900 kilograms of it to Victoria for testing. All seemed well. A mine that was entered by a tunnel driven straight into the hillside was an easier mine to drain, ventilate, and remove coal from, and it was much cheaper to drive a tunnel than to sink a shaft. A writer to the *Colonist* approved of the man in charge: "The energetic manager of Harewood, I believe, will favour a liberal and progressive policy from the first, and advocates the idea of paying men not so much for the time they kill as for the work they do."

But the energetic manager was about to learn a lesson he would use in later life. For a mine to be successful it had to be able to get its coal to market. In the nineteenth century, the only way for an island coal mine to get its coal to market was to transport it to the shore and ship it on oceangoing vessels.

The Harewood lease lay inland behind the VCML lease which blocked access to the two harbours – Nanaimo and Departure Bay – where ships could anchor safely and take on coal easily. The obvious solution was to allow the Harewood Company a rail corridor through the VCML lease and some frontage on Departure Bay for a wharf.

All through 1866 the parties involved argued and proposed deals. The colonial government had ordered the VCML to provide access. The VCML General

Manager, Mr. C.S. Nicol, wanted to force the new company to ship their coal through Nanaimo to increase the value of the town lots the company had put up for sale at inflated prices. Rumours flew. The Harewood mine was about to open. The Harewood mine would never begin operations.

Dunsmuir and the men working for him were to be paid on the basis of the amount of coal they produced, but the coal had to reach market in order to be sold. Since there was no way to get the coal to market, no one was making any money. Even when the VCML agreed to allow a corridor, the company could not afford to build a railway and a wharf. Because of a depression in Britain and the unsure prospects of the Harewood company, the owners could raise no more money from British investors. The eighteen-month deadline was looming.

Realizing it was prudent to leave, Dunsmuir quit the company and returned to the VCML. But he took away with him some lessons to add to the ones he had learned with the HBC: A coal mine was useless unless there was a way to get the coal to market. But if a company could find a way to get the coal to market, there were young aristocrats serving in the Royal Navy at Esquimalt with family money to invest in a well-run mining operation.

All the while he was working for the Harewood Coal Company, Dunsmuir and his family continued to live in their Nanaimo home near the Douglas Pit. Robert had bought the land the house sat on when the company put building lots on the market. A row of trees on the front and side of the lot provided shade and a settled air to the house, where children continued to be born on a regular basis. Jessie Sophie arrived when her mother was thirty-eight years old and Annie Euphemia (Effie) two years later.

While Joan cared for her new daughters, her eldest daughter, Elizabeth, who was now twenty, married John Bryden. The thirty-six-year-old Scot had come to Nanaimo in 1863 and eventually succeeded his father-in-law as superintendent of the VCML. The two men were of like mind, each respecting the other's abilities and each recognizing the importance of service to their community.

With only six hundred permanent residents, Nanaimo depended upon its most able citizens to show leadership in all manner of ways. Bryden and Dunsmuir served together on a committee to find a site for a cemetery and a colonial school. While himself avoiding elected office, Dunsmuir supported political candidates and the campaign to incorporate Nanaimo; he served as reviser of real estate, foreman of a jury, spokesman for Nanaimo's MLA; and he was a founding member of the Total Abstinence Society.

Dunsmuir advocated total abstinence from alcohol for the citizens of Nanaimo but considered it his right to drink alcohol himself. When he worked for the HBC he bought rum from the company store on a regular basis. But like many able men of his time, Robert Dunsmuir believed in one law for himself and another for lesser mortals.

Robert Dunsmuir

As a miner and a mine manager, Dunsmuir had dealt with many kinds of men: the skilled and the unskilled, the angry and the compliant, Canadians, Americans, Aboriginals, Englishmen, and Scots. But into a community chronically short of labour, a new kind of worker had arrived.

In the wake of a six-month-long strike, the VCML had hired Chinese workers for the first time. These men had first come to North America to build railroads in the United States. They came as single men determined to earn enough money to return to China and buy land. The American railroads built, the Chinese came north to British Columbia for the Fraser River and Cariboo gold rushes. When the gold fizzled out, they came to the Island looking for jobs. They were hard workers, but the white miners mistrusted them because they spoke a strange language, wore strange clothes and pigtails, lived separate from the rest of the town, and accepted lower wages.

The first few Chinese worked above ground, but the company quickly saw the advantage of sending them below ground as runners. John Bryden was impressed with how quickly they learned and began hiring them as miners and paying them the same rates as other miners until General Manager Nicol objected. "Why this won't do," he said, and went on in the racist language of the time, "if we are to pay the Chinamen the same rate of wage as the white men there will be no use employing them."

There were over two thousand Chinese in the united colonies of British Columbia and Vancouver Island at the time, and they had been viewed with hostility from the day they arrived. In 1860 a Nanaimo coal miner went to jail for shaking a stick at a Chinese man in Victoria. Seven years later, the *Colonist* observed, "Considerable excitement, we hear, exists at Nanaimo in consequence of the introduction of Chinese labourers. The colliers threaten with violence the first Chinaman who forgets his celestial origin so far as to descend to the 'bottomless pit' of a coal mine." But the Chinese stayed and became one of the ingredients in the explosive mixture of people working in the mines of Vancouver Island.

Another "ingredient" was the strikebreakers who came to Nanaimo during the same strike. The word "strikebreaker" or "scab" had been used in Nanaimo before but only by the Staffordshire men to describe the miners who refused to desert to Bellingham Bay. These new men – unsuccessful American gold miners, impoverished Mexican labourers – were the real thing: men imported from outside to take the jobs of striking miners. Their arrival was sufficient to break the resolve of the strikers, who many thought had been justified in demanding higher wages.

And about this time in Nanaimo, another ingredient to the labour mix had arrived: British union men. By the 1860s unions had become legal in Great Britain, although they had not yet in British Columbia. But because Britain was experiencing an economic recession, there were too many workers for the available jobs. British unions had begun to encourage their members to accept offers of free passage to the New World from North American mine owners.

Among the British miners were men who had been blacklisted and were no longer able to get work at home. Some of these men, experienced miners who knew what unions could do for them, came to Nanaimo just when Robert Dunsmuir was about to increase the demand for labour beyond anyone's imagination.

A proper tweed suit and a chin curtain beard give the new Superintendent of the Harewood Coal Company a respectable appearance.

5

The Managing Partner

There are two things Jimmie Hamilton likes to do. One is fish and the other is shoot the breeze with his cronies at Harvey's General Store and Post Office. Robert Dunsmuir likes to come in to Harvey's too, and Harvey is glad to see him. Dunsmuir's second-eldest daughter, Aggie, has caught Harvey's eye and the handsome and hard-drinking former coal miner is set on marrying her.

Sometimes, Hamilton and Dunsmuir leave the store together and drive out to Diver Lake, an insignificant body of water north of Nanaimo that teems with fish and where Hamilton has his cabin. The two men are a strange match: the kind-hearted Hamilton who loves to fish and the hard-driving Dunsmuir who burns

Robert Dunsmuir

with the desire to catch a different sort of prize. Hamilton knows that when they get to Diver Lake, his friend will disappear into the bush to do what he's been doing for years: search for signs of coal.

One day in October, 1869, the two men climb into Dunsmuir's horse-drawn buggy for another of their expeditions. The road they travel along is really nothing more than a pathway that leads eventually to the settlement of Comox, 120 kilometres to the north. No more than four metres wide at its widest, with dense bush encroaching on either side, the Comox Trail dips and bumps wherever rain-swollen streams have cut across it. Only a determined man like Dunsmuir could keep a buggy from coming to grief on this excuse for a road.

Diver Lake lies just beyond the boundary of the VCML coal lease. Out there, anyone can pre-empt land for settlement or stake a mineral claim. Over the preceding decade, the colonial government has passed legislation to encourage exploration for coal, and it seems to be working. George Robinson has found coal at Nootka Sound on the west coast of the island; John Muir just outside Victoria in Sooke. There is coal at Cowichan, Quatsino Sound, North Saanich, and Comox.

Government regulations allow individuals to explore for coal if they apply for a two-year prospecting licence on 202 hectares. A ten-man partnership can apply for five times as much. But Robert Dunsmuir is not interested in having partners – he has spent the last forty-four years answering to someone else, and he has had enough of it. He leaves Hamilton to fish and disappears into the wilderness of trees and rock to the west of Diver Lake. He will worry about the need for partners when he finds his own coal.

The terrain gradually slopes upward until, about a kilometre or so from the lake, he finds himself on the top of a broad ridge of conglomerate rock that looks down on the west to a lush river valley. The conglomerate ridge excites him. It seems to be the same formation that lies over the Newcastle seam at the head of Commercial Inlet. He is so sure there is coal lying beneath the ridge that he heads back to Nanaimo to tell Joan he has found coal.

The next day he takes another friend back to the site. William Isbister is a stonemason by trade, but he has also worked in the mines with Dunsmuir and knows the strata as well as any man. When Isbister sees the ridge he agrees there is probably coal under the conglomerate rock, but he can't understand what his friend is so excited about. Many people have found coal all over the Island, but no one, not even Robert Dunsmuir, has enough money to develop it.

Undeterred, Dunsmuir hires two miners to help him prospect. They are looking for dirt that is rich and black or for an outcropping of coal that will confirm the presence of a seam beneath the ridge. The three searchers advance northward. One day, two days. On the third they discover a seam of coal over a metre in thickness tipping downward in a southeasterly direction.

In early November, Robert Dunsmuir boarded the screw steamer *Sir James Douglas* for her regular run from Nanaimo to Victoria. Although the ship had been named in honour of the recently knighted former Chief Factor and Governor, she was not worthy of her illustrious name. There were passengers who said she was nothing more than a tugboat. Others complained they had either to remain on the open deck at the mercy of the weather or "be stowed away in the hold, like herrings in a barrel, without any means whatever of obtaining ventilation."

The ship usually made the trip in twelve hours, but on this particular voyage she ran into a gale, forcing her to take shelter in Cadboro Bay, just north of Victoria. Two or three of her passengers, Dunsmuir among them, were too impatient to wait for the storm to pass. They rowed ashore and walked the six kilometres to the colony's capital. Dunsmuir's business would not wait for a mere storm.

Considering the number of people who had found coal in the past decade, the headline in the November 13th edition of the *Colonist* was unusually excited: "A GREAT DISCOVERY of coal has been made at Departure Bay ... The seam is five or six feet thick and crops out on the shore at deep water in a most excellent harbour. Mr. Dunsmuir has secured the lead, so we hear."

Thus, in an announcement whose facts were wrong in almost every respect, the newspaper trumpeted the beginning of the Dunsmuir fortune. Dunsmuir was much less sure than the *Colonist* was, however. He had found a seam of coal, he had the experience and knowledge necessary to develop it, and new legislation required the VCML to allow him shipping access through its lease at Departure Bay. But the amount of land he was allowed to work on was far too small, and the amount of money he would require to prove the claim within the required two years was far too large.

Desperation drove Dunsmuir to get money from whatever source he could. The short-term loan he negotiated with moneylenders in San Francisco had a very high rate of interest. The long-term investment of a young Royal Navy lieutenant named Wadham Neston Diggle meant Dunsmuir had to add Diggle's name to the company's. Thus, having reluctantly taken a partner, he set out to develop a coal mine.

Throughout the spring and summer of 1870, using a crew of five men who agreed to work for future wages, Dunsmuir became better acquainted with the coal he had found. He realized he had not found the Newcastle seam after all, but rather a new one, which he called the Wellington seam. The crew ran a slope down eighty-nine metres into the seam and removed 454 tonnes of coal, twenty-three tonnes of which the men set aside for steam trials to be conducted on the Royal Navy gunboat HMS *Boxer*, which waited at anchor in Departure Bay.

The corridor through the VCML lease was narrow and the terrain steep. Haulage animals – horses and mules – pulling the heavily loaded wagons from the mine inched down a rudimentary road to the beach where open barges or lighters waited in the shallow water. Once loaded, the lighters came alongside the ships at anchor in the deeper water of the bay.

Robert Dunsmuir

The *Boxer* had also taken on coal from the VCML's Douglas and Fitzwilliam mines. In the steam trials that took place in September, Dunsmuir's coal performed the best of the three.

So impressed was the Royal Navy, it wanted to buy all the coal Dunsmuir could produce. But a better way of getting the coal from the mine to the ships had to be devised, and there was no money for railroads and wharves or for more equipment for that matter. In December 1870 Dunsmuir wrote to Joseph Trutch, Chief Commissioner of Lands and Works, to tell him he had tried unsuccessfully to convince his former employer, the Harewood Coal Company, to loan him some boring rods. Now he asked Trutch if the government would lend him some, and two or three chisels and any other small articles that could be spared.

Dunsmuir was desperate. He was still searching for the best place to develop his mine, and he had had to suspend operations more than once due to a lack of funds. So far he had been able to make do without the purchase of much expensive equipment because the coal was relatively close to the surface, but that would not last much longer. Crews were continuing to explore: sinking bore holes, digging pits, discovering coal in various locations. Soon Dunsmuir, Diggle & Company would have to develop an underground mine.

In July, 1871, Dunsmuir discovered the place where the seam was at its richest and where he would sink his slope. As is true for many legendary events, the details of this discovery have become clouded. The historical record contains several versions, including one told by Dunsmuir himself just ten weeks after it occurred, another told by Dunsmuir thirteen years later to Queen Victoria's son-in-law, and a third version told by the daughter of a bootlegger who lived near Diver Lake.

The first version appeared in a letter Dunsmuir wrote to the Honorable H.L. Langevin, Canada's Minister of Public Works, on September 20, 1871. He had once again been "strolling" through the bush. "I chanced upon a root of a fallen tree which I thought had a peculiar appearance. On examination I found coal sticking on the upturned root."

The fallen tree is crucial to each of the stories. The Marquis of Lorne, Canada's Governor General and the husband of Queen Victoria's daughter, Princess Louise, was visiting Vancouver Island in 1882 when he met Robert Dunsmuir and examined the Nanaimo and Wellington coal operations. According to the regal gentleman, Dunsmuir had discovered his coal when he placed a small shot of powder to blast away the surface where he had seen "some slight indication of what he was looking for." Dunsmuir and his companion, "a negro attendant," then walked away into the dense wood to wait for the charge to ignite. Wandering farther than he intended, Dunsmuir "fell in the thicket over the trunk-roots of an uptorn pine. In rising again he grasped at the soil on the roots, and found that his hands had become blackened."

Tilly Smallbones did not tell her version so politely. She said Dunsmuir drank every night at a small-town saloon near Diver Lake, and every night he walked home over a trail which passed through her family's forty-hectare plot of land by the lake and by her family's

Robert Dunsmuir

house. "One night he was so drunk he fell down and went to sleep. During the night he got kicking around and in the morning when he woke up sober found he'd kicked all the moss away and there was a coal mine right on our land."

Although Tilly and her parents, a laundress and a bootlegger of some renown, did live near Diver Lake, her story had a few flaws. It is true that Dunsmuir was known for his drunken revels in drinking establishments within his coal lease. But in 1871, though there was a saloon at the lake, there was no small town anywhere nearby. Since Dunsmuir lived in Nanaimo, nine kilometres away, it is more likely that he would have been driving his buggy home than walking.

Tilly's story concluded with the information that her father had been wanting to sell his land and return to Oregon. She says Dunsmuir bought the land without telling her father there was coal on it. Whenever good fortune happens to someone, there are always people who claim to have been cheated. The Smallbones story sounds like one of those claims.

No one will ever know the exact story of how Dunsmuir found coal in July, 1871, but he did sink a 550-metre slope shortly afterward. When the slope was finished, it was fitted out with a ventilation furnace and haulage machinery. To pay for all this, Dunsmuir had to take on partners. They provided him with the money, and their names on the company documents gave him the right to apply for more land to prospect and eventually to buy.

In order to retain control, Dunsmuir assembled as obedient a group of partners as possible. His two sons, twenty-year-old Jimmie and eighteen-year-old Alex, and his new son-in-law, James Harvey, were obvious choices. They had no money but they would not interfere with his decisions. Diggle had already proven to be easy to manage, content to let Dunsmuir spend his money in the hope of future rewards.

Rear Admiral Arthur Farquhar was the Commander of the Pacific Fleet. So admiring was he of Dunsmuir's abilities that he agreed to invest money and asked four of his flagship officers to lend their names to fulfill the government requirements for a ten-man partnership. The arrangement gave Dunsmuir the land and the money he needed and still left him in control. There may have been ten names on the company documents but there was only one managing partner.

∞

Forty men, among them seven Chinese and eight Indians, had been working for Dunsmuir when the partnership came into being in November 1871.[1] With new money to spend and a new mine to develop, Dunsmuir began hiring more workers. But there was still a shortage of labour on Vancouver Island.

While the VCML always tried to hire family men, Dunsmuir would take anyone who would work: sailors unhappy with the sea, labourers down on their luck, second-generation miners, first-generation immigrants, Englishmen, Scots, Chinese, and Italians. Being of the opinion that the most important thing about a man was

1. In contrast to the 1850s, few native men were interested in coal mining. They did not like to work underground.

how hard he could work, he did not even require that they speak English or know anything about mining.

He hired all comers: a few family men and a lot of single men whose only responsibility was their next meal and their next mug of beer. He gave them jobs, paid them well, and expected complete loyalty. The *Nanaimo Free Press* was very enthusiastic: "Mr. Robert Dunsmuir is a gentleman most admirably adapted for his post," wrote the editor, "although he is exacting from all his hands. There is no man on this coast who is more respected or [well] thought of than he is by every man who is in his employ."

Since most people had to travel on foot, it was necessary for Dunsmuir's men to live within easy walking distance of the mine. He planned at first to offer building lots for sale, but then he changed his mind. Sooner or later the coal would run out and the mine would close. When the mine closed, the town would die, leaving lot owners with worthless land. Dunsmuir knew what was best for his employees; it was better that he provide company houses.

The village of Wellington began to take form near the pithead of Number One Slope: company houses for the families, boarding houses for the single men, a general store, and a bakery.

Two hotels were already in business down the road on Diver Lake. Charles "Donnybrook" Chantrell was a former Nanaimo miner who could smell an opportunity almost as quickly as Robert Dunsmuir could. His house at the north end of the lake had been a place where a man could find a drink for almost as long as Dunsmuir had known about Wellington coal.

Laird of the Mines

Like its rival, The Cosmopolitan, Chantrell's hotel was on private land, but Dunsmuir had to give approval before the government would award him a liquor licence. Akenhead's Boarding House, Akenhead's Wholesale and Retail and Mahrer's Bakery might be privately owned businesses too, but their owners earned a living in Wellington because Dunsmuir allowed them to do so.

Having reaped the benefits of a partnership, Dunsmuir quickly dissolved it, keeping as partners only Diggle and Farquahar, the two men who had loaned him money. His sons and his son-in-law and the four naval officers relinquished their interest within three months of having signed on.

∞

The summer of 1872 would be remembered in Wellington for two things. The first was the constant presence of smoke in the air from the bush fires that burned from Nanaimo as far north as Qualicum. The second was the wooden railway that Dunsmuir built to get his coal to Departure Bay.

From the Admiralty, Diggle had bought two traction engines that became locomotives with the addition of flanged wheels. Over a wooden track built of fir four-by-fours capped with scrap iron, one of the engines brought loaded coal cars from the mine to the top of the steep hill leading down to the bay; the other engine picked them up at the bottom of the hill and took them to the wharves. To get the cars from the top to the bottom, Dunsmuir built a gravity incline. The loaded coal cars were hooked to a continuous wire rope, "the thickest in this country," and lowered down the incline, their weight providing the force to pull the emptied cars up again.

Five years later, two small saddle tank engines named *Duke* and *Duchess* replaced the traction engines. By then, the wharves had been enlarged to accommodate the biggest steamers belonging to the Pacific Mail Steamship Company, a new customer. Dunsmuir had built a company office near the wharves with a large public room and several private offices fitted out with polished walnut.

The office was by far the grandest building at Departure Bay. Humble shanties and sheds clustered behind and beside it. Down the beach to the south was Harper's Departure Bay House, a saloon without a licence to peddle liquor, the government having refused it due to a lack of policemen. Sailors from the ships waiting in the bay to take on coal had nothing to do but loll around on the beach and provoke each other into fighting.

Having no apparent wish to live among the boistrous sailors at Departure Bay or his own employees at Wellington, Robert Dunsmuir continued to live in Nanaimo. In 1872 he and Joan commissioned a contractor to tear down their house near the Douglas Pit and build them something on the same site more in keeping with their new status.

According to the *Colonist*, whose local informants kept a careful eye on the Dunsmuirs, "Ardoon" was a large and beautiful dwelling house. Its two stories and steeply pitched roof nestled into the large lot which sloped upward toward the rear where the stables housed the family horses and buggies.

At forty-four years of age, Joan Dunsmuir had given birth to her last child, Henrietta Maude, that same year. Although her two eldest daughters were married and her eldest son, James, was in Virginia studying mining engineering, Joan still had seven children at home, including Alex. The energetic nineteen-year-old was his father's contact man in various real estate deals and a staunch supporter of such political causes as the incorporation of Nanaimo as a city.

Joan's strong will and influence over her husband and her family were not apparent to outsiders watching the family's fortunes rise. But inside Ardoon her family knew that she had a vigorous mind and a forceful personality and that she played an important part in her husband's business decisions. She was his only real partner. When he left Ardoon each morning with his fast horse and buggy to speed toward Wellington and the day's business, it was her strength that went with him and her refusal to allow him to back down that made him the effective manager he was.

The road he travelled was no longer a trail, and the man he had travelled it with in 1869 was no longer alive. While Robert Dunsmuir had been struggling to prove and develop his coal find through the spring and summer of 1870, Jimmy Hamilton had been content to live at his cabin near Diver Lake and fish whenever he could.

Hamilton's neighbours were the Smallbones. One day in September, in the same month that the Royal Navy was proving the worth of Wellington coal in the

Robert Dunsmuir

Boxer steam trials, two native men called at Smallbones's groggery and, finding no one home, broke in, stole several items, drank all the beer, and left three empty glasses and a naked footprint on the table. As they made their noisy and drunken way past Hamilton's cabin, he came out to protest. In drunken anger, they hit him on the head, cut his throat, dragged him into the cabin, and set it on fire.

Their trial was the talk of Nanaimo that fall, and the Royal Navy sent the gunship HMS *Sparrowhawk* to remain in the harbour until the hanging took place. It happened just a year after Hamilton and Dunsmuir had ridden the Comox Trail to Diver Lake.

By 1876, that trail was a proper but primitive road with small logs laid in corduroy sections to fill the hollows cut by flooding streams. British Columbia had been a province since 1871, and the government had struck a deal with Robert Dunsmuir. The province would build the Wellington Road, as it was to be called, as far as Dunsmuir's boundary. Dunsmuir would complete the construction from the boundary to the village of Wellington. Anyone wishing to work or visit in Wellington was welcome to use the road, but there was a gate at the boundary, an open gate to be sure, but one that Robert Dunsmuir could order closed whenever he deemed it necessary.

In 1875, the pithead of Wellington Number One Slope gives little indication of the wealth it will produce or the lives it will claim.

B.C. Archives 93352

6

Bloodshed Among Us

The ventilation furnace of the Wellington Colliery's Number One Slope is ferociously efficient. Its roaring fire sucks enormous amounts of air through the tunnels of the new mine, creating a strong current that sweeps explosive firedamp from every stall and heading.

The fire in the ventilation furnace requires attention twenty-four hours a day from the three Chinese men – one for each shift – hired to tend it. Chinese make up one third of the work force in Vancouver Island mines in 1876. It has not taken long for white miners to realize that they can hire a Chinese man as a helper, pay him lower wages, produce twice as much coal, and keep the extra profits. And since mine owners are the same as miners in their desire to make more money, they too hire Chinese or "celestial colliers," as the newspapers call them, to be runners and furnace men.

But even though everyone is making money by employing them, no one likes the Chinese being there. The very thing that makes them desirable employees – their willingness to work for lower wages – makes them a threat to other working people. Gangs of thugs in Nanaimo raid Chinese wash houses or break into Chinese stores and hurl boxes of tea and crockery into the street. Police raid their gambling houses, and magistrates penalize their wrongdoers unfairly.

Not surprisingly, the Chinese prefer to live apart from the rest of the community. In Wellington, their dirt-floored shacks form a straggling line to the west of the Number One Slope. Nearby, but separate, is a cluster of company family houses. To the east of the mine over a small bridge that spans a ravine are the village's boarding houses and businesses and, with its back to the road, a twelve-room, two-story house which contains an examining room for the miners' doctor and the home and office of the mine manager.

In the mine manager's office is a wondrous new device: the first telephone on Vancouver Island. Made of yeast cake boxes and powder keg banding, it is the invention of several of Dunsmuir's employees and is used to communicate with Departure Bay and Nanaimo. As useful as it is, however, it will make no one rich, because Alexander Graham Bell has patented his telephone in the United States a few weeks ahead of all his competitors.

Laird of the Mines

The most important thing in Wellington is the mine, its three-story-high chimney and the low white buildings surrounded by scruffy fir trees, fallen logs, and slash. Inside the engine house, mounted on heavy timbers, is the hoisting gear with a hauling rope that stretches down the slope leading into the mine. A straight row of timbers divides the slope into two passages: one side for the hoisting rope and the other for miners entering or leaving the workings.

In the mine on the evening of October 14, 1876, the Chinese man tending the ventilation furnace decides he doesn't want to stand all night in the glare of the hot flames. He piles enough coal onto the fire to last until morning and walks away.

Some hours later, the mine whistle breaks the midnight silence with a fearsome howl. Fire is creeping up the walls of the slope and down into the tunnels and crosscuts of the mine, where there is unlimited fuel to feed it. No one knows how far into the mine the flames extend or how fast they are moving.

Next morning, the ground in front of the colliery office caves in and flames shoot up from the mine beneath. They rise higher and higher until they form a column one hundred metres tall, which roars like a gigantic version of the ventilation furnace that has caused the fire in the first place.

Men work feverishly to seal off the slope and the return air shaft. Four days later, flames no longer shoot into the air but the fire still burns inside the mine. The *Nanaimo Free Press*, however, sees no further cause for alarm: "The proprietors are sanguine that in a few days the fire will be extinguished."

Robert Dunsmuir

However sanguine the proprietors said they were, the column of flames was symbolic of the troubled relationship developing between Dunsmuir and his employees. The heady days when everyone identified with the exciting new enterprise were over. Petty grievances were piling up. Miners were talking among themselves and holding meetings. Then news came that there was too much coal in San Francisco where the most important markets were for Vancouver Island coal.

In July 1876, Dunsmuir had asked his employees to take a reduction of twenty cents per ton from the $1.20 he was paying them. He had to cut the price of coal in order to preserve his markets in San Francisco. When the miners refused his request he laid seventy of them off and kept thirty-six men on to work at the old rate. "You will give me as much coal as I want," Dunsmuir had told the thirty-six, "and you can go to work at $1.20 and the rest can play."

In the ensuing days, deputations of laid-off miners came to see Dunsmuir. They appeared in the middle of the night; they accosted him on the railway track. They demanded they be taken back and paid the $1.20 they were getting before. Dunsmuir held firm; he called for men who would sign a contract agreeing to a fixed sum of money for a fixed amount of coal. The laid-off miners published a "Notice to all Miners" in the *Free Press* advising them "to abstain from coming to [Wellington Colliery] in search of work." But when miners desperate for work offered to dig coal as cheaply as the men on contract, the rest of the men came back to work at one dollar per ton.

If there had been a union, the miners might have insisted on a fairer solution. But though unions had been legal in Canada since 1872, and though there were men in Nanaimo holding meetings to discuss the formation of one, there were no coal miners' unions on Vancouver Island yet.

Shortly after the lay-off ended and just before the furnace fire, James Dunsmuir, twenty-five years old and newly returned to the Island following his July wedding in North Carolina to Laura Surles, took over as mine manager. It was surprising that Robert Dunsmuir, a man who hated to share control, had placed such a young man in charge of the rapidly growing mine. James had worked in the VCML mines as a sixteen-year-old and had trained as a machinist and mining engineer, but he had no experience as a manager. His shortcomings became apparent when the scales at the pithead started to behave crazily.

The amount of coal the miners dug determined how much money they made. When they complained to Robert Dunsmuir that the scales were weighing light, he told them that though he could find nothing wrong with the scales, he would make up any extra the men thought they had lost. Then Dunsmuir asked James to have the scales checked.

The weighman discovered the scales made no sense when weighing anything over 180 kilograms. But James refused any offers from the men to help solve the problem. New scales, he said, could not be ordered because they would take two or three months to come from Britain. In any case, repairs must wait until the superintendent had recovered from an illness. Then an American miner named George Vipond, whom Robert had helped financially, showed James how the upper and lower plates of the scales rested on each other improperly. James finally called a blacksmith who repaired the scales temporarily.

A few days later, a miner named Haggerty complained about the scales again and proposed a solution. Take a few cars down to the wharves and weigh them there. The miners would move them at their own expense. James refused and told Haggerty to pick up his tools – Haggerty was fired.

The Dunsmuirs had not been maintaining the scales at the pithead as well as they should have been. Another complaint finally convinced Robert and James that the scales needed attention. They gave Alex the job of going to Chantrell's and telling the men the scales would be fixed.

Charles "Donnybrook" Chantrell had gotten around to building a proper hotel in 1875 and had given it a proper name. "The Wellington Inn" was a popular place to drink, but everyone still called it Chantrell's. As Alex approached the saloon, he had to pick his way through the menagerie of dogs, pigeons, hens, turkeys, and geese that Chantrell allowed to run free around the place. Alex had lived among miners all his young life, having been the first white baby born in Nanaimo, but when he walked into Chantrell's to tell the miners the good news, they pushed him back outside.

Laird of the Mines

The miners had been meeting regularly at Chantrell's to discuss their grievances. The fires of indignation which had been smouldering since the miners lost the twenty cents per ton had been stoked by the inaccuracy of the scales, and they burst into flame over an old complaint about blasting powder.

Miners used blasting powder, bought from the company, to loosen the coal so they could shovel it into the cars. Wellington coal required an unusual amount of powder – sometimes over a kilogram per shot – to loosen it. The summer before, in a conciliatory gesture, Dunsmuir had agreed to take fifty cents per keg off the blasting powder, but seven months later he still had done nothing about it.

On February 3, 1877, a four-man delegation paid Robert, James, and Alex a visit. This was no ordinary delegation. The men represented the first coal miners' union on Vancouver Island – the newly formed Coalminers' Mutual Protective Society.

The four men presented the Dunsmuirs with an ultimatum: fix the scales and restore the old wages or they would lead the miners out on strike. Since Robert Dunsmuir thought he had been fair, he refused their demands. Two days later, while he was riding down to Departure Bay, he heard that one hundred miners had walked out of the mine.

He took action quickly. He gave the occupants of the company houses, striking miners and their families, one month to vacate his property. He sent Alex to San Francisco to hire new workers, and he published his soon-to-be famous "card" in the *Free Press*.

> There is an impression in the community that we are obliged to accede to the miners' demands: but for the benefit of those whom it may concern we wish to state publicly that we have no intention to ask any of them to work for us again at any price.
> Signed Dunsmuir, Diggle & Company.

The only way a person in the nineteenth century could make his opinion public was through the newspapers. In what would become a lifelong practice, Robert Dunsmuir began to use the newspapers as a weapon. He inserted the "card" regularly over the next four months; he listed the wages of individual miners; he explained and rationalized his various actions – all for public consumption.

A week after his card first appeared in the newspaper Dunsmuir stood at the end of his wharf in Departure Bay to greet the steamer *Etta White* as she docked and disgorged thirty-three new employees – Italians, Englishmen, Frenchmen, and Irishmen – from San Francisco. Dunsmuir welcomed them to the island. The seam they would be working was an excellent one, he told them, and he would provide the tools, although miners usually brought their own. They would sleep in the shanties near the wharf, and by the way, the miners who normally worked this coal were out on strike.

The striking miners, reasoning that no right-thinking man would want to be a strikebreaker, were most anxious to talk to the newcomers. They offered to buy them a drink that Friday evening, but only a few Irishmen accepted. So the strikers were waiting at the

Laird of the Mines

mine the next morning when the company train brought the new men up from the Bay.

Robert Dunsmuir was there too. In intellect and influence, he was a powerful man, but physically he was slight, weighing only 150 pounds and standing much shorter than the average man of his times. As the newcomers walked down the slope into the mine, he stepped in behind them. The strikers followed, their tempers flaring. Dunsmuir turned to face his angry former employees and held his arms out wide to prevent them from going any farther. They retreated to wait outside.

Escorted by the little man, the strikebreakers continued into the mine and stayed there for an hour. When they re-emerged, company men hustled them through the crowd of strikers, who called out to them the reasons for the strike. As the newcomers climbed aboard the train cars for the return trip to Departure Bay, some of them shouted that if they were given money for their expenses they would leave.

Sunday began early in the shanties at Departure Bay. A crowd of strikers roused the San Francisco men from their sleep and escorted them – some reluctant, some cooperative – by foot from Departure Bay to Nanaimo. James tried to intercept them on Departure Bay Road, haranguing them from his buggy to no avail. In Nanaimo, while Robert paced the street outside, the two groups of men met in a Nanaimo hotel where "the chief thing done was drinking." Having convinced some of the men to leave, the union men took them to Victoria by canoe and gave them each fifteen dollars.

It was time for Dunsmuir to enlist the help of his friends in high places. Captain Warren Spalding, Nanaimo's stipendiary magistrate, had been a mounted policeman in Australia and a cavalryman in the Crimea, but now he was in charge of maintaining law and order in Nanaimo. In the small world of Vancouver Island, however, men in power had overlapping responsibilities, and many of them were friends of Robert Dunsmuir.[1]

On February 25th Spalding received a visit from Lieutenant Diggle, who carried a letter from Dunsmuir asking for protection for the new men who had chosen to stay and work. With the letter was another one written by Dunsmuir to the attorney general of the province. "For goodness sake," he had written, "act promptly in this matter, I am afraid that there will be bloodshed among us at this time."

Spalding sat down and wrote a letter to the lieutenant-governor: "The miners are determined to proceed to extremities and have already committed such acts of violence as will necessitate the employment of an armed force to reduce them to order and compel them to respect the law." He placed Dunsmuir's letters in the same envelope and gave them to Diggle to take to Victoria.

Nothing had happened in Wellington and Departure Bay to justify Dunsmuir's frantic plea for prompt action. No blood had been shed; no equipment

1. Five present and future premiers were involved in some way in the strike. Andrew Elliott, George Walkem, Theodore Davie, James Dunsmuir, and Edward Gawlor Prior, who had just started his term as the first mine inspector.

destroyed. But Dunsmuir was determined to prevent even the possibility of such things happening.

The lieutenant-governor's response was quick. Two days after Spalding requested armed force, he received a letter telling him that the government would be sending HMS *Rocket* to render assistance to officers of the law and asking Spalding to inform Dunsmuir and the miners.

Dunsmuir could hardly contain his anxiety. He wrote to the attorney general asking him to hurry. "Such a lot of men I never had to deal with before, and there will be no peace with them until they get a proper lesson, and in haste."

During this anxious time, while everyone waited for the eviction notices on the company houses to expire, a series of union meetings and court trials provided fodder for the newspapers. Who said what to whom and the details of everyone's business, most especially Robert Dunsmuir's, became regular reading material for the public.

Waiting impatiently for the *Rocket* to appear, Dunsmuir wrote another fretful letter to the Attorney General. "We are going to have trouble if not bloodshed when we commence to eject the miners from the houses.... They say they will not leave, and that we cannot put them out."

Finally, the HMS *Rocket* sailed into Departure Bay. Though she had armed marines aboard her, her reputation as "decidedly the reverse of handy" had preceded her. A wit in Victoria predicted that if she were called upon to fire her guns, the shaking would cause her to sink.

Robert Dunsmuir

Whether or not the *Rocket* was a real threat, her very presence was an insult. The miners of Wellington could not believe their eyes. They had gone on strike because they believed it to be the only way they could get "redress of certain deep and urgent wrongs affecting their liberties as men...." The presence of the *Rocket* and its armed men on board was offensive to law-abiding citizens who so far had attempted no breach of the peace.

Other law-abiding citizens were about to feel Dunsmuir's wrath. George Norris was the owner and publisher of the *Nanaimo Free Press* and its only reporter. In that capacity he had attended the first meeting of the new miners' union and published an article about it. One of the miners called Dunsmuir a liar, Norris wrote, and another accused him of knowing in advance that the scales were inaccurate. Dunsmuir sued Norris for libel.

Then he went after the people in the company houses who were ignoring the eviction notices. With ten days to go before the notices expired, Dunsmuir ordered the gate at the border of his property on Wellington Road closed. He told the proprietor of the Red House store in Nanaimo, Alex Mayer, who had been supplying the miners' families with cheaper groceries than they could buy in Wellington, to refrain from bringing in any more groceries. The butcher should kill no beef; the baker fetch no bread. The miners paid Dr. Daniel Cluness a monthly subscription for his medical attention. Dunsmuir gave Cluness fair warning. "As my men have ceased to work, I have no control over their usual monthly payment for [the] doctor and [will] not be responsible for same."

Dunsmuir's miners prefer to work with fish oil lamps even though they cause explosions. The fireboss carries a safety lamp which is designed to detect deadly gas.

B.C. Archives 4245, Ray Knight Collection

7

Sunday Soldiers

At the pithead of Wellington Number One Slope the only sounds on the morning of Thursday, March 8, 1877 are the faint shouts of children playing near the company houses. The eviction deadline has arrived and nothing has happened. Robert Dunsmuir is sitting in Captain Spalding's Nanaimo courtroom this morning listening to striking miners, company officials, and experts of various kinds testify about the accuracy of his scales. It seems as though everyone has an opinion about how he should run his colliery.

Chief Justice Sir Matthew Baillie Begbie waits in Victoria for the chance to add his opinion. He has agreed to begin an inquiry into the dispute between Dunsmuir and his employees "provided that both parties will agree to accept and abide by his decision." Former Premier and now Leader of the Opposition, George Walkem, will represent the miners at the inquiry. Since he and Premier Elliott are political enemies who wage war on each other both in the legislature and in the newspapers, many people question Walkem's motives for representing the miners.

The next day, a member of the union committee meets Dunsmuir by chance in Mead's Barber Shop and discovers that the embattled mine owner has no intention of submitting the question of wages to Begbie's inquiry. On being informed of this, George Walkem advises his clients to withdraw their acceptance.

Thirty-six new strikebreakers from San Francisco have been waiting in Victoria for the signal to come to Wellington. Dunsmuir has asked Premier Elliott to allow the Superintendent of Police to accompany them. His letter to Elliott reveals that though he has expected Begbie's inquiry, he hasn't seen the need of it.

"You must bear in mind," writes the impatient man, "that I would not allow anyone to arbitrate on our business as I can manage that myself, and in fact there is nothing to arbitrate on excepting the breaking of the law by the miners."

Begbie's commission having died before it is even born, there is nothing to stop the new batch of San Francisco men from coming up from Victoria to work in the mine. The *Maude*, a side-wheel steamer with a reputation for repeatedly going aground, delivers the men to Departure Bay, this time without incident. They disembark and board the company train for

Laird of the Mines

Wellington. As requested by Mr. Dunsmuir, Superintendent of Police Todd is in attendance.

The scene at the pithead is unnerving. Women, with babies in their arms and children holding on to their skirts, greet the new men with the question, "Have you come to take the bread and butter out of our mouths?" The new men are easy to convince; soon they are in Chantrell's mingling with the strikers "as though they had been friends for years."

Nine days later, on March 21st, Dunsmuir is back in Spalding's court asking for and receiving possession orders on twelve company houses. A week later a deputy sheriff goes to Wellington to take formal possession. The residents, having consulted George Walkem, who tells them there is no law in the country to compel them to leave their houses, refuse to move.

"We are at last in a fix," writes Dunsmuir to Elliott, "[we] cannot get possession of our property and the law is set at naught. Harris must send force, none can be got here."

Sheriff Harris and his posse of four arrive in Departure Bay on April 3rd aboard the *Rocket*. When they reach Wellington, seventy or eighty miners watch as Harris attempts to take possession of two houses. When his deputies refuse to help him, he sends a note back to the *Rocket* asking for volunteers from among the marines on board. Twelve "blue jackets" answer the call and assist the sheriff in the eviction of two families. The marines then retreat to Departure Bay and scramble aboard the *Rocket*, which steams down Newcastle Channel to Gordon's Wharf in Nanaimo. There Sheriff

Harris disembarks to the hoots and groans of a crowd onshore.

Followed by the noisy throng, Harris walks to the office of lawyer Theodore Davie, where Robert Dunsmuir waits. Bunched outside, the crowd begin to sing a rendition of the "Death of Nelson" to the beating of coal oil tins. Someone cries, "Three cheers and three groans for the brave men who turned the women and cripples out of the houses."

A second attempt three days later to repossess the houses at Wellington meets with another refusal. There is nothing Sheriff Harris can do but exchange "hard" language and withdraw.

If exchanging hard language was the worst thing the miners did, they still would have been within their legal rights. But by refusing to recognize the possession orders they have finally defied the law, and the sheriff is determined to enforce it. When Harris leaves for Victoria, he promises to return with a stronger force.

∞

Since his arrival in the New World, Robert Dunsmuir had enjoyed varying degrees of popularity, but since he had discovered coal, his popularity had known no bounds. Just three years before, a letter writer to the *Free Press* had been rhapsodic. "The people of Nanaimo can look Mr. Dunsmuir in the face and say they feel proud of him. He started on a small capital and through his industry and sagacity, he is today a credit to himself and an honour to Nanaimo."

Then he had reduced wages, thus giving his men a reason to hate him and the new union a cause to fight for. But many citizens, including some of the non-union miners, still admired him. An anonymous "working man" opined that only half a dozen men, all ex-gold miners, had started the strike. According to this writer, everyone else who had joined the strike had done so because they were afraid of being branded "black-legs" or "scabs" as strikebreakers were sometimes called. Other letter writers urged Dunsmuir to send to Pennsylvania, Canada, or even England for more strikebreakers and ignore "this mania for high wages."

But the anti-Dunsmuir forces were growing, and they were out in strength when Sheriff Harris returned to Wellington for the third time accompanied by fourteen young cadets. A crowd of two hundred men and women followed their wagon down the main street of Wellington, past the company office, across the bridge, and up to the cottages at the pithead.

The occupants of the company houses waited behind barricaded doors. The first house after the bridge was Alex Hoggan's. While the crowd watched, Sheriff Harris broke down the door with an axe and burst into the tiny room. Waving a warrant in Hoggan's face, he demanded that Hoggan give up the house. Shouting "damn you and everyone else before you should take anything out of my house," Hoggan lunged at the sheriff, grabbed him around the waist, and pushed him toward the door. Just then, Mrs. Hoggan came out of the bedroom where her child lay sick; she pushed the sheriff against a wall. "This is the dearest day's work you ever did in your life," shouted Harris, shaking with indignation.

Leaving two helpers behind to complete the eviction, Harris came outside to see the crowd pummeling the other young constables with fists and pelting them with sticks and stones. To the sound of oaths and profanity, the sheriff abandoned his adolescent posse and beat a retreat to the relative safety of the colliery office.

The crowd followed, herding the young cadets with them. They threw one young man off the bridge. A striker offered to fight any of the cadets for a hundred dollars. A line of women – "Amazons" according to the *Colonist* – kicked and punched the cadets as they passed by.

That evening in Nanaimo, over a keg of beer, the sheriff and his young posse nursed their bruises and assessed their day's work: three houses sealed, six families evicted – their work was only half done.

Dunsmuir continued to use the newspapers to further his cause. He published the income of fifty miners. The amounts appeared to be high because most of them shared their wages with Chinese helpers. "Not one nail of the houses which they are asked to leave belongs to them," he wrote, "and every man can find land enough on the Island to build a house."

On April 20th, Spalding issued twenty more possession notices and Dunsmuir stepped up the pressure on Elliott. "If the law cannot be carried out, I shall shut down the works for twelve months; and if there is not something done next week I shall do so." As one of the most important employers in the province, Dunsmuir could force the Premier to act.

Laird of the Mines

When Lieutenant-Colonel Charles Frederick Houghton, Deputy Adjutant General of the British Columbia Militia, was a younger man, he had volunteered to lead an exploratory party looking for gold at the head of Okanagan Lake for Governor James Douglas. So badly did he organize and execute the expedition that Douglas concluded he was incompetent. But the passage of years and an ability to be in the right place at the right time had put Houghton in charge of the province's militia, a force which provided military aid to the government against "domestic and foreign foes."

Though only thirty-seven years old in 1877, the Colonel had developed a slight paunch, which lent him an established air in keeping with his lofty rank. Muttonchop side whiskers – narrow at the ear, wide below the chin – adorned his face and a handlebar mustache completed the stylish effect. When he was in full regalia, the extravagant white feather on his hat off-set his facial plumage magnificently.

Communities all over the province maintained militia units where civilians could play at being soldiers, marching and shooting their guns in their spare time in return for the chance to "aid the civil power" in times of crisis. Mobilizing the militia required a letter bearing the signatures of three justices of the peace. Houghton received such a letter on April 23, 1877. It was signed by three Victoria JPs and Nanaimo's Captain Spalding. Robert Dunsmuir had lost patience waiting for Premier Elliott to do his bidding and had collected the signatures himself.

Robert Dunsmuir

The letter spoke of the situation in Wellington and of the likelihood of a riot or disturbance of the peace when the sheriff again attempted to evict the miners. It asked Houghton to call out as much of the active militia as he considered necessary. Since the Nanaimo unit was small and singularly inept, Houghton would have to bring in soldiers from outside.[1]

"The strikers number about one hundred," he wrote in his letterbook, "and a very large majority of the men of the other coal mines in the vicinity [are] known to sympathize with them." He was pleased when the muster of Victoria and New Westminster produced fifty-six men. Five officers and fifty-one Sunday soldiers – bank clerks and butchers, law clerks and haberdashers – filled their haversacks with food, a blanket, a mattress, and twenty rounds of ammunition and notified their employers that they would be absent while they protected the people of Wellington from harm.

It was 10:00 p.m. on Sunday April 29th, when the *Maude*, crammed with soldiers and bound for Departure Bay, set sail into a strong wind out of the northwest. All night the steamer rolled and pitched in the heavy headwind and delivered a tired and seasick cargo at 9:30 the following morning. Robert Dunsmuir, his top hat giving him added height, waited on the wharf with Magistrate Spalding at his side. Having spruced up their uniforms and rubbed the shine back into their boots, the militia disembarked and marched

1. According to Houghton, "There is not a single officer or man in the [Nanaimo] corps who knows anything whatever with regard to drill much less is capable of giving instruction therein."

to the foot of the gravity incline where coal-blackened wagons awaited them – their transportation into battle.

Crushed into the dirty cars, the soldiers endured the journey up the incline, the hitching of the engine at the top and the lurching of the cars over the uneven rail bed to the pithead. For a battleground, Wellington was strangely quiet. The mining families had just been to the funeral of a small boy who had drowned in a well. But as the militia climbed out of the coal cars and stretched their cramped legs, the atmosphere lightened. Catcalls and good-hearted joking from the crowd put the soldiers at their ease.

"My God, what's them?"

"Ain't there a lot of 'em too."

"A dose or two of buckshot would clear 'em all out."

"They've come to eat turkey till they bust."

But for all the banter, the militia's mission was a serious one. The soldiers were there to protect the sheriff while he evicted these people from their homes. As the crowd watched and some women shouted abuse against the Dunsmuirs, the sheriff proceeded. Some families went quietly, others begged to be allowed to stay, one or two shouted obscenities as the Sheriff dragged them away.

It was difficult work. By 4:00 p.m. only five families had been ejected; three miners had been arrested and household goods from the empty houses carried to the train to be taken off company property. When a rumour spread that the Nanaimo union men were on their way to help the strikers, Magistrate Spalding panicked and ordered Number One Company Victoria Rifles to guard the property over night.

Robert Dunsmuir

While the majority of the soldiers slept on the *Maude* with sentries posted to prevent them from visiting the pubs of Nanaimo or their prisoners from escaping, double sentries stood guard on each of the three most important pithead buildings. All night the calls of "wild animals" kept the sentries at the mine awake. In the morning the militia again watched as the sheriff completed the evictions. A woman wished out loud for some rotten eggs to throw. "If times were not so hard," she said, "I would have thrown good ones."

The behaviour of the soldiers was exemplary as twenty families, three or four for the second time, became homeless under their watchful eyes. They returned to the *Maude* to sleep and sailed from Departure Bay the following evening.

There was no moon that night. True to form, the *Maude* ran aground on a rock. Though she did not take on water, she was unable to proceed. There was nothing to do but wait through the night and hope for rescue in the morning. Since the crowded conditions made it impossible for anyone to lie down, it was a tired and dishevelled lot that welcomed the sight of a passing steamer the following day.

Once back on dry land, Houghton declared himself to be delighted with the performance of the militia. "There can be no doubt of the fact that were it not for the presence of the Militia at Wellington, the miners would have continued to set the law at defiance."

Robert Dunsmuir was pleased too. When the strike was over he invited Houghton to visit his family at Ardoon, where the Colonel first laid eyes on Marion Dunsmuir, fifteen years his junior. They were married in 1879 and sailed away on the *Maude* for their honeymoon.

⁓

In the weeks that followed the militia's visit to Wellington Dunsmuir brought in more strikebreakers. This time the city of Victoria provided two policemen to accompany them, the law officers' salaries being paid by the provincial government. The strikebreakers went to work unopposed. Six miners stood trial for "unlawful combination to raise the rate of wages" but five of the six were found innocent. Alex Hoggan, the one man found guilty, went to jail for four months for obstructing the sheriff.

The Wellington mine was open again, manned by "the most ragged, forbidding lot of men that ever set foot in the Province, and not one has ever been in a coal mine in his life." With the addition of yet another shipment of men, Dunsmuir could run all three shifts again. When forty local men, all former employees of the Wellington mine, went to the colliery office and asked for work, Dunsmuir told them he did not require their services. He had said repeatedly during the strike that no one who joined a union would ever work for him again, and Robert Dunsmuir always meant what he said.

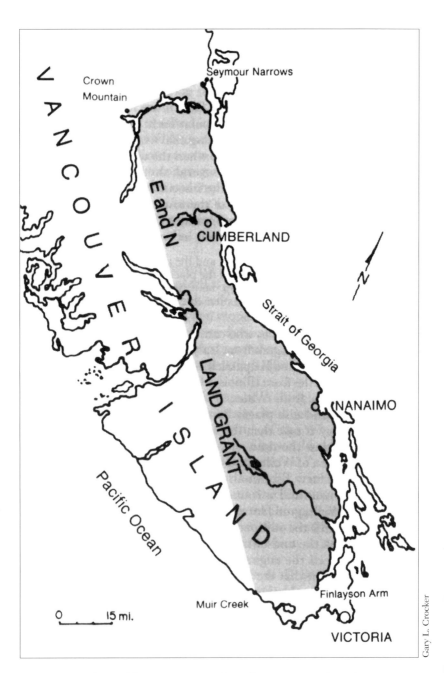

In return for building the Esquimalt and Nanaimo Railway (E & N), Dunsmuir receives an 800,000-hectare land grant.

8

"In Full Everyday Working Blast"

Reuben Gough was no scab. He was born in Nanaimo to a mining family – one that came from Staffordshire on the *Princess Royal* – and his father, Edwin, was working to establish a union when he lost his leg in a mining accident. That was when Reuben was six years old. His mother, Elizabeth, like many women whose husbands were disabled or killed in the mines, became the family breadwinner. She operated a hotel on Commercial Street in Nanaimo, where she sold liquor and rented rooms to miners.

Reuben's father died in 1875, and his mother's hotel burned down three years later. Elizabeth needed someone to support the family while she rebuilt. Since fourteen-year-old Reuben was the oldest boy still at home, he was the logical candidate to quit school and go into the mines.

Fourteen-year-old boys had been going into the mines for generations in Britain. Until the 1840s, boys and girls much younger than fourteen had been used in British mines to haul coal out of low places on their hands and knees. But in British Columbia, the 1877 Coal Mines Regulations Act (CMR Act), prohibits boys under twelve from working underground. Fortunately for the Goughs, Reuben was old enough to mine coal. He got a job in the Number One Slope at Wellington.

∽

Dunsmuir is not happy about the CMR Act. He has been writing his own rules for some time now, and he does not like being told what he can and can't do. He and other mine managers resent the rules that say they must not allow boys under sixteen to work underground more than six hours a day, that miners should be allowed to appoint their own check weighman, that no owner or manager can act as an inspector, that an inspector can shut down a dangerous mine, and "that no Chinese shall be employed in any position where his neglect or carelessness might endanger the limbs or lives of any man working in the mines."

The belief that Chinese are dangerous has been growing as their numbers in the mines increase. The furnace fire in Wellington just before the strike has confirmed people's suspicions. But as production increases dramatically in the months after the strike,

Laird of the Mines

Dunsmuir needs more employees, and the Chinese are there, ready to work and for lower wages, too.

Dunsmuir still needs experienced miners as well. The need for labour is so acute that he has taken back all the union men he swore he would never employ again. By 1879, even Alex Hoggan is working at Wellington again.

As production increases, so does the presence of firedamp. Every day, all through Dunsmuir's mine, firedamp is bleeding out of the newly disturbed coal. The newest level, Number Ten, has been gassy from the beginning, but the furnace-driven ventilation is so effective that no one worries much about an explosion. Even when a fire ignites in a stall worked by a man named Horne on the evening of April 15, 1879, no one is very worried.

But by the middle of the night the fire is still not under control. Dunsmuir himself takes charge, shouting orders, scolding miners, directing crews, and ordering the installation of curtains to direct the air properly. When he finally emerges from the mine twelve hours later, he has left behind a sign written in large chalk letters on the curtain by Number Ten level: "NO ONE IS ALLOWED TO PASS HERE – FIRE."

But there are many men in the mine, including the hundreds of Chinese, who can not speak or read English. Dunsmuir has said the Chinese are as safe as anyone else, but he cautions the shift foremen to make sure no one goes behind the curtain. All through the evening and into the night, crews keep the coal wet and the fire under control. No one, including the foreman who usually checks for gas, goes behind the curtain.

Robert Dunsmuir

All foremen carry safety lamps whose protected flames warn of gas but will not ignite it. But safety lamps are unpopular: miners think them too heavy and their lights too dim; mine owners think them too expensive. In the Wellington mine that night, as the gas accumulates behind the curtain in Horne's heading, every man has a naked-flame lamp hooked to his cap and burning brightly.

With his Clanny safety lamp in his hand, John Dixon, the day shift foreman, lights his oil lamp and relieves the night foreman at six o'clock on the morning of Thursday, April 17th. Behind him the night foreman hears the supervisor of the Chinese runners send two men down to a miner who is grumbling about having no empty cars. Dixon heads down to Number Ten level shortly afterwards. Thirty minutes later the mine whistle screams its hoarse call over and over and over.

There has been an explosion. Heedless of the danger of carbon monoxide – or afterdamp – that follows a mine explosion, the inspector races down the slope accompanied by John Bryden, manager of the VCML mines and Dunsmuir's son-in-law. From Number Seven level down, there is evidence of a "fearful blast." At the mouth of Number Ten level they find "a perfect chaos": coal cars piled in a heap, the roof caved in, steel rails torn up and twisted. Outside the curtain with the warning message are several bodies lying in heaps like discarded toys. Inside are four more: John Dixon and three Chinese.

By now miners and managers from all over the area have come to help. Braving the chokedamp they re-establish the flow of air and carry the dead away.

Two days later the coal face at Horne's heading is still so hot that the water hitting it comes away nearly boiling. Steam and smoke fill the air. Seeing no alternative, and over the furious objections of Dunsmuir and Bryden, the inspector orders the mine flooded. Crews divert a small river into the mine.

One man's body is still inside: a Chinese known only by his employment number. The inquest finds that it was he who caused the explosion by going behind the curtain. But the inspector warns that as long as naked flame lamps continue to be used in Vancouver Island's mines, "disaster is sure to follow."

Families and friends prepare for the burial of three Chinese, three Englishmen, one Italian, one Irishman, one American, and fourteen-year-old Reuben Gough. The eleventh man's body will not be recovered until the water is pumped out of the mine nine months later.

∞

John Bryden envied his father-in-law his freedom to make decisions and see them implemented immediately. Bryden managed the VCML mines but took all his orders from a board of directors in faraway London, England. He was fed up with waiting for months to hear from his superiors and having his decisions challenged by a group of men who knew little about mining and whose primary concern was making more money for the shareholders.

Robert Dunsmuir had to answer only to himself, but he was unhappy with the way his son was running the mine. James had too little experience and he lacked his father's ability to deal with men on their own level. He was young, stiff-necked, and unsure of himself. Bryden was mature, experienced, and very able. The senior Dunsmuir courted him, visiting the VCML office frequently and taking him away for fishing trips when the pressure of work became too much.

In 1881 Bryden replaced James as manager and moved to Wellington with his family. James and Laura Dunsmuir moved to Departure Bay, where James was to be in charge of the wharves. Their new house with its pillared porch stood in splendid isolation nearby. Their only neighbours were the Chinese who worked on the wharves and the hundreds of sailors who patronized Joseph Harper's Saloon, which now had a liquor licence, while they waited in rowdy boredom for their ships to be loaded. After an evening of drunken fun – stampeding horses and firing revolvers into the air – several of the "foreign and migratory element" usually ended up in Bummer's Hall, as they called the local jail. Sunrise would often reveal a body lying dead on the beach.

James's younger brother, Alexander, was living in San Francisco selling coal and hiring men for Dunsmuir, Diggle & Company. Gossips reported that he had moved into the apartment of a recently divorced woman named Josephine Wallace. Officially she was his landlady; unofficially she was his mistress. But Alex's unseemly love life and his fondness for alcohol were unknown to his parents; he was his father's most successful son.

With his sons working for the company, his son-in-law in charge of his mine, and profits climbing rapidly, Dunsmuir planned a triumphant return to Scotland.

Thirty years before, he and Joan had slipped out of Kilmarnock quietly on very short notice; now they would return affluent and proud, accompanied by their unmarried daughters. But even while he planned the extensive journey, including stops in Ottawa, London, Glasgow, Paris and returning across the United States to San Francisco, Dunsmuir was making plans for when he came home.

On February 18, 1882, at a farewell dinner in his honour, he announced that he would run for a seat in the provincial parliament. The fact that the election was in July and he would not be returning from his trip until September did not concern him. His sons-in-law, John Bryden and James Harvey, would campaign in his absence.

That Robert Dunsmuir won the election was a surprise to some. Not only was he absent from the province but ever since the 1877 strike he was no longer as popular with the local citizenry as he had been. And since the rule requiring voters to be property owners had recently been dropped, there were many who hoped that the newly franchised working class would vote for a working class candidate.[1] But these new voters were still unsure of their power and voted instead for establishment candidates. When the votes were counted on July 24th, Dunsmuir won by a small margin: 226 votes out of 424 cast.

---

1. For a person to qualify as a voter he had to be a man and he had to be over twenty-one and a Canadian or a British citizen who had been in the province for at least a year and in the constituency for two months.

Robert Dunsmuir

The new Member of the Provincial Parliament (MPP) for Nanaimo had been making plans to move to Victoria long before he won the election. A man of his wealth and importance should live in the most important city in the province. And his wife needed to be where their five unmarried daughters were more likely to find suitable husbands.

Victoria had shed its fur-trade-gold-rush image and had taken on an established and genteel air with two-story brick buildings on the main streets and grand Victorian houses on the quiet residential boulevards. With its gleaming white exterior and distinctive bay windows, "Fairview," as the Dunsmuir's new home was called, was the grandest of all. Its location near the harbour and across from the legislature made it convenient for the new MPP and highly visible to the citizens of Victoria, who wondered how the newly wealthy family would fit in to Victoria society.

In the thirteen years since he had discovered coal, Dunsmuir had acquired property and wealth at a remarkable rate. In 1878 he had bought out two of his three partners: Admiral Farquahar and a Captain Egerton, who had provided cash in 1873. In 1883 he bought out Wadham Diggle, paying him $600,000 for his $8,000 original investment, and changed the company's name to Robert Dunsmuir and Sons. His Wellington Colliery Railway had five locomotives, 150 wagons, and ten miles of track. He was a major shareholder in the Canadian Pacific Navigation Company, and had just bought the Albion Iron Works. He owned a fleet of eight colliers and a magnificent steam-driven tug named the *Alexander*.

The *Alexander* was an incongruous sight on Friday, October 20, 1882 when she steamed into Departure Bay, not towing a sailing ship as she customarily did, but decorated from bow to stern with flags and bunting. Even her bow wave seemed festive as she steamed toward the colliery wharf where Robert Dunsmuir waited to join her illustrious passenger – the Marquis of Lorne, Governor General of Canada – for a short voyage to Nanaimo.

No one as important as Queen Victoria's son-in-law had ever come to Nanaimo before. Evergreen boughs, Chinese lanterns, and flags too numerous to count decorated the streets. As the vice-regal parade proceeded up Commercial Street and onto Victoria Crescent, it passed under no fewer than five arches installed and decorated for the glorious occasion. Speeches and visits to VCML mines filled the day before the Governor General was able to retreat to the Dunsmuirs' home just off the Crescent, where Joan and her staff had prepared accommodations for him for the night.

At the entrance to Ardoon was yet another arch, this one topped with a crown and flanked by two Union Jacks. An inner arch coloured in shades of red, white, and blue led up the porch steps under a "Welcome" sign. Evergreen boughs festooned the stair and balcony railings; flags draped the balcony, the arch, and several flagpoles. Inside, the entire Wellington Brass Band waited to play for the honoured guest.

When the Governor General emerged from Ardoon at nine o'clock that evening to attend the

Robert Dunsmuir

Citizens' Ball, two large reflecting gas lights illuminated the arch at Dunsmuir's door. To the Marquis's right, up Victoria Crescent and opposite the Hong Hang & Co. store, the residents of Chinatown had spent $750, more than twice a Chinese miner's yearly wage, to build their square arch with three pagoda roofs. Wax figures and flags, glowing lanterns, and hanging bouquets of silk and paper flowers reminded the Queen's representative not to forget Nanaimo's least popular citizens.

At the Citizens' Ball, the Marquis apologized for the absence of his wife, the Princess Louise, who was indisposed in Victoria and whose absence had disappointed many. Then he danced a Scottish reel with the mayor's wife before retiring early in order to prepare for a visit the next day to his host's mines at Wellington.

The visit to Wellington had been publicized as a chance for the Marquis to see a colliery "in full everyday working blast." But sometime on that Saturday as the carriage carrying the Marquis and the coal baron travelled over the bumps and winds of Wellington Road, or as they swept by the hundred or so small white company houses that now crowded closely together along Wellington's main street, or as they examined the huge sample of coal outside the office or the loading facilities at the mine or the wharves at Departure Bay, the aristocrat and the industrialist discussed business.

The Marquis had summoned Dunsmuir to Government House in Victoria a few days before. The two men walked in the garden for three hours talking about the building of a railway from Esquimalt, outside

Laird of the Mines

Victoria, to Nanaimo. During that garden walk and now on the tour of Wellington, Dunsmuir had to pretend that the whole idea was a surprise to him.

In truth, Dunsmuir had been discussing a railway with Canadian cabinet minister Sir Charles Tupper and Prime Minister Sir John A. Macdonald for over a year by then. Dunsmuir was interested in the land grant and mineral rights that went with building the railway; Tupper and Macdonald were interested in appeasing the British Columbia electorate, which had been promised a transcontinental railway when they agreed to become part of Canada in 1871.

No one knew in 1871 what route the railway would take through the British Columbia mountains or where its terminus would be, but as the two-year deadline for the start of construction loomed, Macdonald decreed that the terminus would be in Esquimalt. During the next decade, as Macdonald's party lost and regained power and scandals threatened to sabotage the whole railway project, the people of British Columbia waited impatiently. In 1878 George Walkem, the miners' lawyer during the 1877 strike, became premier of British Columbia again and Macdonald again became prime minister, this time as Victoria's Member of Parliament.

By now, the Canadian Pacific Railway (CPR) had chosen the Fraser Valley as the route and Burrard Inlet as the terminus for the transcontinental railway. But Macdonald's own constituents had not forgotten his promise that the railway would end in Esquimalt. The only man with enough money, ability, and influence to build the 112 kilometres of railway that would satisfy

Robert Dunsmuir

their demands was Robert Dunsmuir. Since the reward for doing the job was a strip of land thirty-two kilometres wide on either side of the railway from Victoria to Seymour Narrows and a $750,000 federal contribution toward construction costs, Dunsmuir was very interested in doing his patriotic duty.

The Settlement Act of 1884 – so called because it settled the railway question – granted the principals of the Esquimalt and Nanaimo Railway (E & N) – Robert and James Dunsmuir, John Bryden, and four American railway tycoons – the surface rights to 800,000 hectares of the most arable and accessible land on Vancouver Island, the timber on it, and everything below the surface except gold and silver.

Such an agreement called for a celebration in Wellington. John Bryden gave a group of miners fifty dollars to toast the momentous event; the company clerk threw in an extra ten. But a letter to the editor of the *Free Press* said that 99 per cent of the local people were opposed to the Act and called those who were celebrating "blacklegs and nobsticks." There had been another union-inspired strike the previous year, and the outrage people felt toward the man who would not allow them to have a union would not be soothed by sixty dollars worth of free beer.

Dunsmuir, the future builder of the E & N Railway, stands in top hat beside his locomotive, the *Duchess*, at Departure Bay.

B.C. Archives HP49566

9

A Princely Fortune

At fifty-nine years of age, Robert Dunsmuir has the white hair and receding hairline of an elder statesman and the upright posture of a man used to wielding power. But he does not look like a typical late-nineteenth-century tycoon: no rotund belly pushes against his waistcoat, no close-cropped beard covers his chin and cheeks. Instead, he is slim and his face is clean-shaven except for a spare "chin curtain" beard that clings to his jawline. He looks like a man who has lived carefully and avoided excess.

Rumours of an excessive love of alcohol persist, however. When he still lived in Nanaimo, so the story goes, his own miners often had to carry him home after a drinking spree. Other rumours say that he saves his binges for his trips to San Francisco. But wherever he does his drinking, it never impairs his ability to make a deal or utter a cutting remark or defy a determined union organizer.

The most successful capitalist in British Columbia is just as determined as ever to keep unions out of his mines. In 1883, miners from Wellington organize a mass meeting to discuss forming a union. Dunsmuir fires any of his employees who join. The union calls a strike, and on the same day Dunsmuir gives eviction notices to the people living in company houses.

It is 1877 all over again. Dunsmuir orders in special police; he directs Alex to send strikebreakers from San Francisco; he publishes lists of wages and points out that "nearly all the miners at Wellington employ a Chinaman." The *Colonist* reports that he has sent his steamer *Wellington* to Hong Kong with orders to return with "a cargo of coolies." Since the round trip will take three months, it looks as if Dunsmuir thinks the strike will last at least that long or that the men he is importing as strikebreakers will be kept on after the strike.

Realizing that hiring Chinese themselves has weakened their cause, the union men agree unanimously to stop employing Chinese. Dunsmuir's promise to remove Chinese coal diggers from the mine gradually ends the strike. But when Dunsmuir rehires the strikers, it is only as individuals, not as union men. And he hires forty Chinese who had formerly worked for individual miners.

It is another setback for the union, but the impulse to organize has grown too strong for it to die. In January 1884, a formerly secret American society that claims to be "more than a trade union," calls its first meeting in Nanaimo. Preaching self-improvement and temperance, the Knights of Labor seek to become a secular religion for the working class.

Knowing that the Settlement Act and its huge land grant will provide the necessary cause to rally their troops around, the Knights of Labor attack Dunsmuir at every opportunity. In their testimony before a royal commission investigating Oriental immigration, they call the Settlement Act "infamous." "All the immensely valuable coal lands contained within the vast railway reserve [have] been handed over to one company," the delegate declares, "the principal shareholder in which but a few years ago, without a dollar [developed] a few acres of coal land, which the then favourable laws of the province allowed him to acquire."

Having admitted that the cost of developing the coal has been high and the profits relatively small, the delegate concludes petulantly, "Yet so huge have been the profits that he has accumulated a princely fortune, and has become all powerful in the province, his influence pervading every part of our provincial government, and threatening its very existence."

Everything that Dunsmuir has done is now seen as a threat to someone. The royal commission has been called because anti-Chinese feelings are rampant province wide, and many citizens want to prevent any further immigration from China.[1] Now the completion

1. Since 90 per cent of Canada's Chinese in the late nineteenth century lived in British Columbia, the federal government was slow to respond to concerns regarding Chinese immigration. When it did respond it was with the infamous head tax, which charged Chinese immigrants first fifty dollars and later five hundred dollars to immigrate.

of the CPR will release hundreds more Chinese onto the British Columbia job market. By hiring Chinese, Dunsmuir appears to be against the majority of British Columbians. As the editor of the *Free Press* wrote during the 1883 strike:

> It is a well known fact that the proprietors of the Wellington Collieries are filling the places of the striking miners with Chinese.... [U]nless some means are taken, either legislative or otherwise, to check this evil, in a few years British Columbians will have to hand over this fair and beautiful province to the heathen Chinese.

But Dunsmuir has not been hiring Chinese out of the goodness of his heart. His only interest is in saving money on wages and, more importantly, in the principle that no one – not the union, not the government, not the people – tells Robert Dunsmuir how to run his empire.

But Dunsmuir is not above telling the people what they want to hear. In October of 1884 he grants the Knights of Labor an interview. He does not like Chinese as diggers or loaders, he says, but he needs them for now as runners. "But as soon as I get another mine open which will enable me to give every miner a place to himself and still produce the necessary amount of coal to fill my orders, I shall then stop the miners hiring Chinamen to load their coal. John cannot dig coal at all, at least not my coal...."[1]

---

1. "John" was a name given to any Chinese. British Columbians seemed unable to remember or understand Chinese names.

Laird of the Mines

It is obvious that neither the miners nor their employer have been true to their promise to exclude Chinese from the mines. Nor is Dunsmuir likely to be. He needs every man he can hire for the new mines he is developing: the Alexandra Mine, which he names for the Princess of Wales, and his biggest mine yet, Number Five Shaft at Diver Lake.

The new mines will avoid the mistakes of the past. An explosion in Wellington's Number Four mine in June, 1884, has killed twenty-three men. The inquest blames a negligent fireboss, open-flame lamps, and too much gas. John Bryden will make Number Five as safe as a mine can be. A large fan runs a ventilation system so strong it blows out miners' lamps.

∽

In Victoria, Robert Dunsmuir had taken the city by storm. The *Colonist* praised his every move and utterance; the *Times* criticized him with equal vigour. The speed with which the E & N was progressing was a constant topic of conversation. The gold and silver coins Dunsmuir used when he paid his bills added handsomely to the economy of the capital city. The citizens heard and saw his name everywhere. Even the steamer that carried passengers from Nanaimo to New Westminster was called the *Robert Dunsmuir*, although the people of Nanaimo who had known the great man in his humbler days and were familiar with the boat's sooty exterior, called it "Dirty Bob."

Dunsmuir had just completed the purchase of an eleven-hectare building lot on a hilltop overlooking

Victoria. Below, in the growing business section, he was about to open a brand new theatre.

Opening night of the Victoria Theatre was a glittering affair as one thousand lavishly dressed patrons stepped down from their carriages and swept in between Corinthian columns that supported an elegant Tuscan portico bathed in a wash of gaslight. At eight o'clock the "brilliant throng" was in its seats; the excited chatter subsided as the curtain rose.

Seated in front of a drop curtain that pictured a castle set among groves of trees and romantic lakes were five men in evening dress – the four directors and the secretary of the Victoria Theatre Company Limited – their prosperous faces illuminated by 110 gas footlights. Seated in the middle was Robert Dunsmuir, the man who held the mortgage.

Dunsmuir rose to speak. "I am sure that never before has such a handsome assemblage graced a theatre." Loud applause. "Had I not taken such a prominent part in forwarding the building," he went on modestly, "I would be unstinted in my praise, but I will instead leave the audience to judge the worth of the magnificent structure." Cheers. After more self-congratulatory comments by the mayor, the party surrendered the stage to an amateur production of *The Pirates of Penzance*.

∞

During the election campaign of 1886, the controversy over the Settlement Act took centre stage. Dunsmuir was standing for re-election in Nanaimo, and that was where the opposition to the act was the loudest. Already, three years before, there had been demands for his resignation.

When asked in 1883 to attend a meeting of the electorate at the Institute Hall in Nanaimo, he had sent instead a printed address, which the secretary had read out loud. No one else wanted to build the E & N, Dunsmuir wrote, so he had taken on the job. "Having put my hands to the plough, I am not the man to turn back. I have the contract and I am going to build the railway and telegraph, and operate them too."

Now, as he fought the 1886 election, the railway was nearing completion. This time several candidates, including two representing labour, opposed Dunsmuir and his running mate, William Raybould. But the day had not yet arrived when a labour candidate could defeat a man with so much power. Dunsmuir and Raybould won the seats, the former getting 366 votes out of a total of 610.[1]

Crews had been building the E & N Railway from both ends of the line. By mid-August 1886, they met in the middle at Shawnigan Lake. Dunsmuir's company had finished the 112-kilometre line through rough terrain and dense forest in two years and four months – well before the agreed upon deadline.

On August 13, 1886, Prime Minister Sir John A. and Lady Macdonald joined Robert and Joan Dunsmuir on board Dunsmuir's private rail coach, *Maude*, for the journey to Shawnigan Lake and the driving of the last spike. Workers had made sure that

1. In a by-election called the next year following Raybould's death, a labour candidate finally won the second seat in Nanaimo.

the ceremony would go flawlessly. They had primed the hole in the wooden tie with soap so the golden spike would slip in easily when the prime minister tapped it with a small silver mallet.

The labours of the day accomplished, the official party proceeded by rail to Nanaimo, where they dined at the Royal Hotel on Commercial Street. After luncheon, to the delight of onlookers, Sir John took the air on the hotel's second-floor balcony.

But Sir John, who shared a love of strong drink with his host, had other things on his mind. He had expressed a fascination for the intricacies of coal mining and now accepted Dunsmuir's offer of a trip down a mine. The two powerful men descended on the cage. Wearing coveralls over their elegant clothes and free of their disapproving wives, they toasted the E & N in proper style with generous amounts of Scotch whisky.

By the next year Dunsmuir had extended the E & N from Nanaimo to Wellington. From there a branch line reached to the pithead of Number Five, where coal for the locomotive steam boilers could be loaded directly into the tenders of each train.

On May 3, 1887, the stakes were raised in the game of gambling the lives of miners for higher coal production on Vancouver Island. In a big new VCML mine whose tunnels reached out under Nanaimo's harbour, an explosion took 147 lives. The cause was "a blowing out shot," which propelled fire out of a drill hole and into a tunnel and ignited large amounts of dry coal dust suspended in the air. The resulting explosion ripped through the entire mine.

"A blowing out shot" occurred when a miner set his blasting powder improperly. The large amount of dry coal dust was a result of a ventilation system that was too powerful. The huge fan that drove the system had been a mistake.

John Bryden spoke at the inquest. "I dread coal dust more than I do gas," he said. But he dismissed the possibility that naked flame lamps were to blame. Coal dust required a long, strong flame to ignite it, the type of flame that occurred when a miner took shortcuts setting his blasting shot. But Bryden was fatalistic. "To make coal mines absolutely safe you would have to close the mine."

The inquest absolved Chinese miners of any responsibility. Though forty-eight of them had died, no Chinese had been anywhere near the origin of the explosion. The surviving miners scoffed. They were sure the Chinese were the cause, and they began to meet to demand that Chinese be excluded totally from the mines.

John Bryden set about making Number Five Wellington mine safe from an explosion like the one in Nanaimo. A system of pipes carried water from Diver Lake to every corner of the mine and kept the coal dust wet. The air in the mine carried a fine spray of water. And Number Five was the best ventilated mine in the district. The large fan often blew out the miners' lamps.

Robert Dunsmuir had been in San Francisco with his wife and two of their daughters when the explosion occurred in the VCML mine. He returned to find that the provincial cabinet was about to propose his name to Queen Victoria for a knighthood.

It all seemed very fitting to Dunsmuir's cronies in government. The Queen would be celebrating her Golden Jubilee and would be honouring the Empire's most illustrious subjects. How fitting that British Columbia's coal baron should have a real title. In the meantime, the government appointed Dunsmuir to a cabinet position as President of the Executive Council without portfolio.[1]

That summer of 1887, while Nanaimo families adjusted to life without their fathers and brothers, and Wellington families looked over their shoulders to see if bad luck would strike again, Robert Dunsmuir began the construction of a castle.

To have a home fit for an aristocrat could have been the reason he decided to begin building his baronial residence. A better reason would have been in gratitude to Joan for her steadfastness and good counsel. If the legend of his promise to build her a castle in the New World was true, then he was finally making good on it.

As architect for the castle he chose an American with a reputation for designing imposing houses for Portland's business elite. The castle would be a reflection of Dunsmuir's idealized memories of Scotland. It would be built of stone as the castles in Scotland were

1. The position is an obscure one that probably involved chairing cabinet meetings and is now part of the Premier's job description.

Laird of the Mines

and would be named Craigdarroch after the mansion of the legendary Annie Laurie. It would have a huge stone fireplace, paneled walls, a ballroom, and a tower with a commanding view. It would be worth ten times more than any other luxury house in British Columbia and hundreds of times more than a miner's cottage.

With a castle to build, a railway to operate, coal mines and foundries and hotels to manage, and a constituency to care for, the powerful Scot was a busy man with much on his mind. In the provincial parliament the Opposition waited for him to make a mistake either during debate or in his business dealings. Newspapers, those in his camp and those outside, reported his every move, quoting him and misquoting him as it suited their own purposes.

Dunsmuir delighted in taking on his enemies, especially during debate in parliament. When the Member for Comox, Thomas Basil Humphreys, asked Dunsmuir when the E & N intended to extend the railway to Comox, Dunsmuir replied. "I, as president of the council, consulted the president of the Esquimalt and Nanaimo Railway company, and the president of that railway company informed me that as president of the council I was to mind my own business. Therefore I cannot answer the honourable member's question."

Undeterred, Humphreys continued to look for evidence that Dunsmuir was misbehaving. On a trip to meet with his sons in Portland, Oregon, in December 1887, Dunsmuir granted an interview to a reporter from the *Portland News*. The next day, the newspaper published an article which said Dunsmuir regretted that Vancouver Island was not a part of the United States.

When Humphreys brought the matter of Dunsmuir's "treasonous statement" to parliament, the debate was lively. But as it wore on, Opposition members grew increasingly less sure of their ground. They refused to serve on a select committee to investigate the alleged treason. The committee, comprised of government members entirely, met once, heard Robert Dunsmuir's version of events, and cleared his name.

Members of the government laughed at Dunsmuir's witty statements in the House, and audiences cheered when he appeared on the political hustings. He was British Columbia's grand old man. But as newspapers in eastern Canada and England repeated Humphrey's charges, the likelihood of Robert Dunsmuir receiving a knighthood became very remote. He would have to content himself with the accolades of his own province.

Members of B.C.'s parliament pose in 1888. In front, from the left, are future premiers Theodore Davie and John Robson, an unidentified man, and Robert Dunsmuir.

B.C. Archives 7748

10

Laird of the Mines

On January 25, 1888, at 8:15 a.m., a reckless miner working in Wellington Number Five mine lights the squib on an improperly set blasting shot. Flame tongues out of the drill hole, ignites the dust that fills the air in the tunnel, and starts a chain of explosions, which roar throughout the new mine.

The blast kills sixty men, but ninety-one survive. They struggle through the wreckage and darkness knowing that lethal carbon monoxide gas or afterdamp is filling the tunnels and crosscuts behind them. They squeeze under the wreckage of fallen timbers and rock; they stumble over the bodies of the dead. When they finally get to shaft bottom they discover they are trapped – the impact has disrupted the hoisting cage and the ventilation system. But above ground, the fan is unharmed; as soon as rescuers fashion a temporary chimney out of canvas, the fresh air begins to flow, gradually removing the afterdamp and the threat to the trapped miners.

The dead are mutilated and burned. Elisha Davis's body has been cut almost in half, his head and his left leg blown off. William Wilks's body is burned. A Belgian miner's body is naked; his clothes are completely incinerated.

An E & N train carrying Robert Dunsmuir and doctors and medical supplies sets a speed record as it careens up from Victoria. The train loses a flange on one of its wheels and barely escapes disaster itself. When it reaches the mine the grizzly job of finding and identifying the dead has already begun.

Helpers wrap victims in canvas and carry them to the temporary morgue in the carpenter shop. Family or friends identify the bodies; volunteers wash, dress, and place each one in a coffin. Thirty-one white miners, eleven Chinese company runners, and eighteen Chinese backhands have died.

Rescuers find Robert Greenwell lying dead beside his backhand, Ah Kee. No one knows anything about Ah Kee, but Greenwell and his two brothers came to the Island from Cape Breton in 1884. His brother, John, was a supporter of Dunsmuir and signed a petition in 1886 asking him to stand for re-election. But after the explosion kills his brother, John joins the Knights of Labor.

Island miners are convinced that the culprits in the two huge explosions of 1887 and 1888 are Chinese.

Laird of the Mines

One week after the sixty men die in Number Five mine, miners and managers come together in a mass meeting in Nanaimo, the appalling loss of life having united them if only temporarily. The E & N offers free train tickets to the Wellington men so they can attend.

It is the first of a series of meetings held to address the terrible tragedy. Everyone knows that change is necessary. They agree quickly to the removal of coal dust, the prevention of blowing out shots and the appointment of trained shotlighters. But the final demand requires several meetings to accomplish: the removal of Chinese from all underground workings.

On February 6th, Sam Robins, superintendent of the VCML, and John Bryden agree to the miners' demands. A sceptic in the crowd shouts that Bryden should have to put his promise in writing. But most of the crowd is euphoric. Three cheers go up for Robins, Bryden, and Dunsmuir.

Robins moves his Chinese to the sorting tables at the VCML pitheads and to a tract of company land to clear it for miners' homesteads. Dunsmuir sends his Chinese to Cumberland near Comox, where he is building a railroad from his new mines to his new loading wharves at Union Bay. Having owned these coal lands for almost a decade, Dunsmuir has finally yielded to his sons' entreaties to begin development. Rumour has it that the Dunsmuirs will allow the Chinese to dig coal in these new mines.

In a debate in the provincial parliament regarding amendments to the Coal Mines Regulations Act, Dunsmuir claims that he heard nothing about removing Chinese from the mines until the Knights of Labor made it their rallying cry. All this talk about unions makes him feel protective toward the men who stay loyal to him. "Genuine miners do not attend the meetings and take no part in the agitation," he says.

The following month the Knights of Labor call an idle day to insure a good turnout at a meeting. Bryden threatens to close the mines or put the Chinese to work. Dunsmuir supports his decision with a telegram. The men back down but one miner sums up their frustration when he says of Dunsmuir, "He has nearly as much power in Victoria as the Czar in Russia. While he is erecting buildings, bridges and railways in Victoria, he is making widows and orphans in Wellington."

∞

The railway and bridge most recently completed drew spectators by the thousands on March 29, 1888, to celebrate the new extension of the E & N from Esquimalt to Victoria. Flags, streamers, and banners floated from storefronts and streetlamps. "Success to R. Dunsmuir, Victoria's best friend" and "The Right Man in the Right Place," the banners proclaimed. Locomotive Number Four pulled the *Maude* and three coaches – all decorated and loaded with dignitaries, guests and hangers-on – across the bridge over the Inner Harbour.

At the station on the other side, a procession of carriages waited to convey the "grand old man" and his son, Alex, through the streets to a banquet in their honour. Many congratulatory speeches awaited the man who had accomplished so much for British Columbia, but as the train was coming across the bridge a figure dressed all in black caught Dunsmuir's eye. It was Dr. Gustavus Hamilton Griffin, a recent arrival in the city, who had tried unsuccessfully to sell Dunsmuir some coal lands and was now speaking ill of him to all who would listen.

As Dunsmuir stepped down from the train and settled himself in his carriage, his employees from the Albion Iron Works unhitched the horses and themselves pulled his carriage to the banquet. It was a familiar nineteenth-century custom, the humble on foot pulling the mighty in carriages, and it was reserved for royalty and men who had done great things.

The great man's castle had been under construction for seven months and required the labour of so many stonemasons that there was often a city-wide shortage of them. The death of its architect in January 1888 had put the project in some peril; the castle was only half completed when two lawsuits diverted Dunsmuir's attention.

The first was the famous Black Hand case. The police had arrested Dr. Griffin and the lawyers had been hired when the *Evening Times* published an article claiming Robert Dunsmuir "carried the government in his breeches pocket." Not only was the government building roads to his mines, the newspaper said, but Dunsmuir was selling land in the railway grant to settlers for the outrageous price of three dollars an acre.

The *Times* had been in the anti-Dunsmuir camp for years and just that spring had published a song that called Dunsmuir "King Grab." But this time the newspaper's accusations were too much. Dunsmuir directed his lawyer, Theodore Davie, who was also the province's attorney general, to sue for libel.

On November 23rd, before His Lordship Sir M.B. Begbie and a special jury, Davie called his client to the witness stand. The only persons paying three dollars an acre were land speculators, Dunsmuir testified; settlers paid one dollar. The government had not paid for the road into the Cumberland mines; he had paid for it himself and built a railway as well.

The cross-examination by defense counsel Mr. Hett showed how Dunsmuir played with his opponents whenever he could.

"What would you value the [E & N] grant at?" asked Mr. Hett.

"Well one hundred years from now I will be better able to answer that question. What will you give?" answered Dunsmuir.

"Would you accept five million dollars?"

"Down with it and I'll tell you."

"Would you accept four million dollars?"

"Down with it and I'll tell you."

"The witness is too Scotch for you," interjected Begbie. "He answers one question with another."

The defense team argued that the newspaper was entitled to make fair comment and if found guilty would not be able to bear a large fine. The jury found the newspaper guilty of libel, and Begbie directed judgment to be entered for five hundred dollars and costs.

It seemed as if Robert Dunsmuir won all his battles. No one – not blackmailers nor journalists nor union organizers – could touch him. His enormous power and his ability to provide jobs for hundreds of people combined with his wit and resilience made him unbeatable.

When the Knights of Labor again threatened his Wellington mines by promising a strike if men weren't paid better for doing company work, Dunsmuir reacted like a weary parent. He shut down the mines and explained that he wanted to save his men from making a mistake. "Some of the miners ... have used bad language about me, and I cannot promise to forgive them all," he said sadly to a delegation of miners from Wellington in his Victoria office on Government Street. "I want you to understand that I have shut down the mines so you could not strike, as I do not want to cast you out these cold days."

It was a cold day in January, 1889. John Bryden was with Dunsmuir as he met with the delegation. The four miners, Arthur Spencer, Joseph Carter, Ambrose Moissant, and John Greenwell, were respectful but determined as they presented their grievances.

Dunsmuir's main concern was the Knights of Labor. He repeated what he had said so many times before when he told them he would not have unions in his mines. "[I]f you do not like that policy you can go elsewhere to find work, though most of you I shall wish to employ again, because most of you are good men when the agitators are not working upon you." Then he asked them why the union had sent a delegation to his new mines in Cumberland.

"I believe it came on because some of them heard the Chinamen were working in the mines at [Cumberland]," replied Spencer, "and it seems to be the truth."

"It does not matter to you if I work Hottentots up there," Dunsmuir scolded, "as long as I do not send them down to the mines in which you are working. What difference is it to you?"

Having accomplished very little, the deputation accepted Dunsmuir's offer of free passage back to Wellington on a special train. As they left, Dunsmuir's final words echoed in their ears: "And when you arrive tell the men that I am a stubborn Scotchman, and that a multitude cannot coerce or drive me."

Stubbornness won the day. When the miners heard the next day that Dunsmuir had told Bryden to empty the company houses, the strikers decided to go back to work.

∞

But the grand old man was not feeling as powerful and determined as he appeared. He confessed to his friends that he had feelings of doom. On the recommendation of a grocer named Mr. Fell, he sought out a spiritualistic medium, who gained his confidence by telling him something about his youngest daughter, Maude. Then the medium told him he would die in April.

Dunsmuir was sixty-four years old but those who had heard him duelling with Mr. Hett in the courts or revelling in the parry and thrust of debate in parliament said he was as vigorous as ever. On April 5th, the day before the legislative session ended, the MPP for New Westminster, James Orr, met Dunsmuir as he was leaving the members' room.

"I will see you again in a short time, Laird," Orr said respectfully, "as I frequently come down to Victoria."

"I don't think I shall ever see you again," Dunsmuir said.

"Why, I will be down here. Are you going to Europe?"

"No. I don't think I'm going to live long."

"That's nonsense. You are good yet for twenty years."

"Na, na, Jamie, I dinna think that."

Within five days, Dunsmuir was sick with a cold. On Wednesday April 10th his family discovered him lying in bed "in a death-like state" and summoned Dr. James Helmcken, who in turn summoned his father, Dr. John Helmcken, and Dr. J. C. Davie.[1] The three medical men were able to restore him to consciousness.

Dunsmuir had prepared a will that divided his empire between his two sons but he had never signed it. An earlier will left everything to Joan. Now he asked the doctors if he was in danger. If he was, he wished to sign his more recent will. The doctors assured him there was no danger.

It appeared the doctors were right. Dunsmuir's condition improved through Thursday but then he

1. James Douglas Helmcken was named for his maternal grandfather, Sir James Douglas. John Helmcken was the same man who was at Fort Rupert just before Dunsmuir arrived there in 1851.

suffered a relapse on Friday, April 12th. All afternoon his condition deteriorated until 6:40 that evening when he died, with Joan and his daughters, Emily and Effie, at his side. His two sons had already been summoned to Victoria. His eldest daughters, Elizabeth Bryden and Agnes Harvey, had arrived with their husbands by special train the night before. Mary Dunsmuir Croft was on the high seas taking her sister Maude to school in England. Marion Houghton was in Montreal.

Dunsmuir's oldest friend, Edward Walker, was at Departure Bay where he heard the news the same day. Since he and Dunsmuir had left Fort Rupert, Walker had lived in Nanaimo and then moved to the Bay with his wife and large family. The jack-of-all-trades had prepared the ground for Dunsmuir's first pithead, helped build the wharves, and worked above ground at Number Three and Four shafts. He and Dunsmuir were the same age and they had both worked hard all their lives, but Walker's life had been on a humbler scale.

The *Colonist* mentioned Dunsmuir's humble friend in the black-bordered columns that announced his death from "an accumulation of uric acid which resulted in blood poisoning." Walker's name reinforced the newspaper's opinion that Dunsmuir "was a kind and sympathetic friend. The good that he did ... will live long and his name will be gratefully and pleasantly remembered by the hundreds whom he has benefited."

The plans for the public funeral filled half a column of the newspaper. The account of it filled two. Tuesday, April 16th began with a viewing at Fairview. A steady line of visitors entered the home to look on

Robert Dunsmuir for the last time. He lay in a white-satin-lined metal casket, his head on a pillow of roses, under a full plate glass front. Flowers filled the room and later covered the casket and filled the hearse during the dignified procession through the black-draped streets of Victoria.

Thousands of citizens and visitors watched from the streets as marching bands, soldiers and sailors, firemen, politicians, and policemen paraded by. The *Nanaimo Free Press* had announced that hundreds of Dunsmuir's employees would pull the hearse, but this funeral was no place for such medieval gestures. Instead, four black horses drew the coach bearing the casket past innumerable flags flying at half-mast. Male mourners followed in carriages; the women by custom stayed at home. Sons, sons-in-law, doctors, lieutenant-governors and former lieutenant-governors, judges, consuls, and bureaucrats – twelve hundred men made up the cortege. The procession took a half hour to pass any given point.

The bells of all the city's churches tolled during the funeral service at St. Andrew's Presbyterian Church. In the years since his death the man they were honouring has been described as "the grim old pioneer of industry," "a hard-boiled Scotchman," a man with "many enemies and many ardent and admiring friends," "the most approachable of men," "practical, level-headed, [a man who] knew what he wanted and took the shortest route."

Robert Dunsmuir lived at a time when the myth of the self-made man was important in Canadian society. Such men had humble origins and were self-reliant, determined, and ingenious. According to the myth, if young people would only read about the lives of these men, they would be inspired to do as well.

Few of the miners who worked in Wellington would have been interested in what made Dunsmuir so successful. But they did give him grudging admiration. Many people could tell stories about his generosity to people in financial trouble. But they also remembered how he treated them like children and how there never was a union in any mine owned by Robert Dunsmuir.

Craigdarroch Castle may have been a promise kept to his wife Joan, but Dunsmuir died before he could live in this symbol of his success.

B.C. Archives 5445

Epilogue

The Dunsmuir empire passed to Joan Dunsmuir. Her two sons had never been given any autonomy by their father, and now his failure to sign the new will that would have made them his heirs robbed them of that autonomy after his death.

When Craigdarroch was completed, the widow Dunsmuir moved in with her three unmarried daughters, and for a time the castle hosted brilliant parties and lavish weddings fit for the daughters and granddaughters of a coal baron.

The family business and the family members' relationships with each other foundered as Alex sank towards an early death from alcoholism and Joan sued James over Alex's will.

Stolid, unimaginative James became the premier of British Columbia for two uninspiring years and then the province's lieutenant-governor. He is remembered, if at all, for his castle, Hatley Park. The third generation of Dunsmuirs were more adept at spending money than making it and squandered the fortune on Paris fashions and Monte Carlo games of chance.

Robert Dunsmuir

After another decade of strikes and mine disasters, the Dunsmuirs shut Wellington down and opened new mines at Extension, south of Nanaimo, where the labour unrest and mine deaths continued. In 1911 they sold their mines to a new company, which retained the family name in the firm's title. By 1927 Canadian Collieries (Dunsmuir) Ltd. owned all the mines on Vancouver Island.

The Chinese continued to be persecuted in British Columbia. It was not until 1947 that they were allowed to immigrate on an equal basis with other new Canadians and to enjoy all the rights of citizens.

The miners didn't get their union until 1937 when most of the mines on the Island had already closed down. In the fight to get Canadian Collieries (Dunsmuir) Ltd. to sign the union agreement, the anger of the workers was often focused on the name "Dunsmuir." The coal miner who had done so well in the New World will be remembered always by the working people of British Columbia as the man who would not allow unions.

Although he became B.C.'s premier and, later, lieutenant-governor, James Dunsmuir was a lacklustre successor to his dynamic father.

Chronology of Robert Dunsmuir (1825-1889)

Dunsmuir and His Times	Canada and the World
	1670 Hudson's Bay Company (HBC) receives royal charter to trade for furs in North America.
	1775 Emancipation Act ends slavery in Scottish coal fields.
	1800 Coal has become a driving force in technological development throughout Western Europe and North America.
	1824 British parliament allows workmen to form combinations or unions for collective bargaining.

Robert Dunsmuir

Dunsmuir and His Times

1825
Robert Dunsmuir is born on August 31 in either Hurlford or Burleith, Ayrshire, Scotland, to James and Elizabeth Hamilton Dunsmore; no record of birth or baptism survives.

1828
Joanna Oliver White (future wife of Dunsmuir) born on July 22 to Alexander and Agnes Crooks White; baptized in Kilmarnock, Ayrshire.

1832
Dunsmuir's parents, two sisters and grandmother die within two weeks of each other.

1835
Dunsmuir's grandfather (his guardian) dies; Dunsmuir becomes the ward of his aunt and uncle, Jean and Boyd Gilmour.

Dunsmuir enrolls in Kilmarnock Academy and later in Paisley Mercantile and Mechanical School.

Canada and the World

1830
Gunpowder has come into use for blasting in coal mines.

1835
Invention of the shaft cage makes it possible to load coal into cars and lift them to the surface.

Coastal Indians report the presence of coal at Beaver Harbour on Vancouver Island to HBC.

1837
Victoria becomes Queen of Great Britain and the Empire.

Rebellions occur in Upper and Lower Canada.

Laird of the Mines

Dunsmuir and His Times	Canada and the World
	1841 Upper and Lower Canada are united.
	1842 British act abolishes the employment of women and children in underground coal mines.
	1846 Oregon Treaty places HBC headquarters at Fort Vancouver in American Territory; HBC headquarters moves to Fort Victoria on Vancouver Island.
1847 Dunsmuir marries Joanna (now Joan) White on September 11; their first child, Elizabeth Hamilton, is born eight days later.	**1847** British Factory Act restricts the working day for women and children (aged thirteen to eighteen) to ten hours.
	1848 Royal Navy begins to change from sail to steam power giving HBC a market for its coal; Muir party is recruited and sails from Ayrshire to Fort Rupert, Vancouver Island.
	California gold rush begins.
	Revolutions erupt in several European countries.
1849 The Dunsmuirs' second child, Agnes Crooks, is born.	**1849** Colony of Vancouver Island is established; Richard Blanshard is appointed Governor.

151

Robert Dunsmuir

Dunsmuir and His Times	Canada and the World
1850	**1850**
HBC recruits the Gilmour party in Ayrshire in November; word reaches Ayrshire of desertions and murders at Fort Rupert.	First coal miners' strike on Vancouver Island in April.
	HBC clerk, Joseph McKay, observes coal at Nanaimo in May.
The Dunsmuirs join the Gilmour party in early December.	Most of the Muir family deserts Fort Rupert in July; three sailors murdered by Nuwitti; Chief Factor James Douglas sends a message to Scotland requesting more miners.
The Gilmour party leaves for London on December 9; sails on the *Pekin* December 19 for the west coast of North America via Cape Horn.	
	California becomes a state of the United States (U.S.)
	School of Mines established in London.
1851	**1851**
After a six-month-long voyage, the *Pekin* runs aground in the mouth of the Columbia River in Oregon Territory.	James Douglas is appointed Governor of the Vancouver Island colony.
The Dunsmuirs' third child, James, is born July 8 in Fort Vancouver, Oregon Territory.	HMS *Daphne* attacks a Nuwitti village on July 9.
The Gilmour party arrives at Fort Rupert on August 9 on board the *Mary Dare*.	
1852	**1852**
Gilmour leaves Fort Rupert for Nanaimo in December leaving Dunsmuir and Edward Walker (Dunsmuir's friend) behind.	HBC commences mining in Nanaimo in September.

Laird of the Mines

DUNSMUIR AND HIS TIMES

1853
The Dunsmuir family moves to Nanaimo in either February or April.

Gilmour finds coal at the head of Commercial Inlet in June.

Dunsmuir proves a second seam at the head of Commercial Inlet.

The Dunsmuirs' fourth child, Alexander, is born.

1854
Nanaimo's Bastion is completed in June; Nanaimo becomes a coaling station for the Royal Navy fighting the Crimean War.

Staffordshire miners arrive in Nanaimo on November 27.

The Gilmour family returns to Scotland.

1855
Staffordshire miners stage a series of desertions and strikes; John Meakin threatens his wife; women and children go into the Bastion while the Kwakiutl and the Sne ney mux settle their differences.

Dunsmuir's HBC contract expires in August; he is granted a free miner's licence on October 12.

The Dunsmuirs' fifth child, Marion, is born.

CANADA AND THE WORLD

1853
Vaccination against smallpox is made compulsory in Britain.

1854
Britain, France, and Turkey declare war on Russia in Crimea.

1855
Pend d'Orielle River gold rush begins in Washington Territory.

153

Robert Dunsmuir

DUNSMUIR AND HIS TIMES	CANADA AND THE WORLD
1856 Dunsmuir begins to work the Level Free Mine as an independent contractor on February 29; by April, all available men are working for Dunsmuir, albeit reluctantly.	
A grieving and distraught mine manager, George Robinson, hits Oversman John McGregor on the head with a hammer and deals with Dunsmuir unfairly.	
1858 Robinson returns to England.	**1858** Ottawa is chosen capital of Canada.
Dunsmuir builds a larger home on the corner of Albert and Wallace Streets in Nanaimo.	In British Columbia (B.C.), the Fraser River gold rush begins and attracts large numbers of men who relieve the labour shortage in the colony; Chinese men are among the immigrants.
	The mainland colony of B.C. is established; James Douglas becomes Governor of the new colony too.
1860 The HBC opens the Douglas Pit.	
	1861 American Civil War begins.

Laird of the Mines

Dunsmuir and His Times

1862
The HBC sells its Nanaimo possessions to the Vancouver Coal Mining and Land Company (VCML).

The Dunsmuirs' sixth child, Mary Jean, is born.

1863
Dunsmuir becomes the superintendent of the VCML.

Dr. Alfred Benson discovers Harewood coal; he and his partner, Lieutenant-Commander the Honourable Horace Lacelles, form the Harewood Mining Company.

Dunsmuir joins the newly formed Total Abstinence Society.

John Bryden, future Dunsmuir son-in-law, is hired by the VCML as a coal viewer.

1864
Dunsmuir retires from the VCML to become the superintendent of the Harewood Company in April.

Dunsmuir is chairman of the school and cemetery meetings and a meeting to overthrow the Member of the Legislative Assembly (MLA)

The Dunsmuirs' seventh child, Emily Ellen, is born.

Canada and the World

1862
Smallpox epidemic decimates the native populations of Vancouver Island.

1863
James Douglas retires as Governor of B.C. and Vancouver Island and is knighted by Queen Victoria.

Arthur Kennedy is appointed Governor of Vancouver Island.

Robert Dunsmuir

Dunsmuir and His Times

1865
Dunsmuir forms the Committee of Electors, whose candidate for the Legislative Assembly, J.J. Southgate, wins the June election.

1866
Dunsmuir is on the committee to lobby for Nanaimo's incorporation as a city; he is part of the delegation to Governor Kennedy.

The Dunsmuirs' eighth child, Jessie Sophia, is born.

1867
The VCML hires thirty Chinese workers; white miners protest; the company locks them out and the miners strike for two months.

Elizabeth Hamilton Dunsmuir marries John Bryden.

1868
Dunsmuir resigns as the superintendent of the Harewood Coal Company and returns to the VCML.

The Dunsmuirs' ninth child, Annie Euphemia, is born.

Canada and the World

1865
President Lincoln is assassinated; American Civil War ends.

1866
B.C. and Vancouver Island colonies unite; Frederick Seymour is appointed Governor.

1867
Canadian Confederation unites Ontario, Quebec, Nova Scotia, and New Brunswick; John A. Macdonald becomes the first prime minister and is knighted by Queen Victoria.

Das Kapital, Volume I, by Karl Marx, is published.

1868
First regular Trades Union Congress is held in Manchester, England.

Laird of the Mines

DUNSMUIR AND HIS TIMES

1869
Dunsmuir discovers Wellington coal in October; he registers his claim in November and applies for a prospecting licence.

Boyd Gilmour dies in Scotland.

1870
Dunsmuir negotiates with the provincial government for better prospecting terms; he initiates a survey and tries various locations for slopes and shafts; he builds a wagon road to Departure Bay and the wharves.

HMS *Boxer* conducts sea trials comparing VCML and Dunsmuir coal; Dunsmuir coal is best.

Jimmy Hamilton (Dunsmuir's friend) is murdered in September.

CANADA AND THE WORLD

1869
Louis Riel leads the Red River Rebellion.

The linking of two American railroads, Central Pacific and Union Pacific, creates the first North American transcontinental railroad and telegraph and improves travel and communications dramatically for the entire West Coast.

1870
Manitoba joins Canadian Confederation.

Unification of Italy; many Italian workers emigrate to work in North American mines where their desperate financial situation makes them agree to work as strikebreakers.

Robert Dunsmuir

Dunsmuir and His Times

1871
Wadham Neston Diggle becomes Dunsmuir's partner.

Dunsmuir establishes the site of Wellington Number One Slope in September.

Ten-man partnership is formed in November so Dunsmuir can control a larger amount of coal land.

Charles Chantrell offers liquor for sale near the Wellington mine.

Agnes Crooks Dunsmuir marries James Harvey.

1872
Their names no longer necessary on the documents, six of Dunsmuir's partners resign from the company.

Dunsmuir builds a wooden railway and gravity incline to transport coal to the wharves.

Dunsmuir tears down his second home and builds "Ardoon" on the same site.

The Dunsmuirs' tenth and last child, Henrietta Maude, is born.

1873
The village of Wellington is established.

Canada and the World

1871
B.C. joins Canadian Confederation on the promise that a railway will be begun within two years to connect the province to the rest of Canada; the railway terminus is to be at Esquimalt on Vancouver Island.

British parliament legalizes labour unions.

1872
British Ballot Act requires secret balloting.

Canada legalizes trades unions and gives members greater protection.

1873
Sir John A. Macdonald loses the federal election.

Prince Edward Island joins Canadian Confederation.

Laird of the Mines

DUNSMUIR AND HIS TIMES

1874
Dr. Daniel Cluness is appointed the first medical officer for Wellington Colliery.

Diggle buys two traction engines for the mine railway from the Admiralty.

1875
The Wellington Inn (Chantrell's) opens officially after Dunsmuir gives approval for the liquor licence.

Wellington Chinatown is established near Number One Slope.

James Dunsmuir marries Laura Surles.

CANADA AND THE WORLD

1874
Indians and Chinese lose the right to vote in B.C. elections.

Robert Dunsmuir

Dunsmuir and His Times

1876
Dunsmuir lays seventy men off in July when they refuse to take twenty cents less per ton; the miners return to work in August having gained nothing.

James assumes the day-to-day management of colliery shortly after the strike ends.

An over-stoked ventilation furnace causes a mine fire in October; a Chinese worker is blamed.

Miners complain about the accuracy of the scales in November.

The province builds Wellington Road to the gates of Dunsmuir's property; Dunsmuir completes the road to Wellington and installs a gate.

Canada and the World

1876
Alexander Graham Bell registers a patent for the telephone and founds the Bell Telephone Company.

Laird of the Mines

DUNSMUIR AND HIS TIMES

1877
Wellington miners strike on February 3; Dunsmuir gives one-month eviction notices on company houses on February 8; on February 26, the Coalminers' Mutual Protective Society is founded and Dunsmuir closes gate on the Wellington Road; between February and June repeated unsuccessful attempts to evict miners results in Lieutenant-Colonel Houghton bringing in the militia; the strike ends with the union unrecognized.

Telegraph built to connect the mine to Departure Bay; first telephone in B.C. connects the Wellington mine and the wharf with Nanaimo.

1878
Alexander is sent to San Francisco to run the company office.

CANADA AND THE WORLD

1877
Sir James Douglas dies in Victoria on August 2.

The B.C. Coal Mines Regulations Act is passed on August 15.

1878
The Honourable George Walkem re-elected Premier of B.C.; Sir John A. Macdonald re-elected Prime Minister of Canada and MP for Victoria (until 1882).

Marquis of Lorne becomes the Governor General of Canada.

Chinese banned from B.C. government employment.

161

Robert Dunsmuir

Dunsmuir and His Times

1879
Dunsmuir buys the South Wellington mines; the Wellington Colliery Railway replaces the primitive one.

An explosion in the Wellington mine kills eleven men on April 17; a Chinese worker is blamed.

Marion Dunsmuir marries Lieutenant-Colonel Charles Frederick Houghton.

1880
Dunsmuir opens several new mines and acquires more coal lands.

1881
John Bryden becomes manager of the Wellington Mines; James is demoted to run the wharves at Departure Bay.

Sir Charles Tupper asks Dunsmuir to present a proposal to the federal government for building the Esquimalt & Nanaimo Railway (E & N).

Explosion in Wellington Number Four Shaft.

Canada and the World

1881
Federation of Organized Trades and Labor Unions of U.S. and Canada is founded.

Laird of the Mines

Dunsmuir and His Times

1882
After announcing his candidacy for election to the provincial legislature, Dunsmuir leaves for a trip to Scotland; he wins his seat in absentia.

Dunsmuir buys the Albion Iron Works, the largest foundry north of San Francisco.

The Marquis of Lorne, Governor-General, visits B.C. and asks Dunsmuir to build a railway.

1883
Dunsmuir moves to Victoria to live in "Fairview"; he buys out Diggle and forms Robert Dunsmuir and Sons.

Reborn Miners' Mutual Protective Association holds a mass meeting.

Dunsmuir signs a railway contract with the federal government.

Wellington miners strike from August to November.

Public meeting in November over the building of the railway demands Dunsmuir's resignation from his seat in the legislature.

Canada and the World

1882
U.S. passes an act to prevent Chinese immigrants entry for ten years.

1883
Marquis of Lorne's appointment as Governor General ends.

Robert Dunsmuir

Dunsmuir and His Times

1884
Knights of Labor establish a local in Nanaimo in January.

The Settlement Act gives Dunsmuir an 800,000-hectare land grant and $750,000 to build the E & N; the first stake is driven May 6.

Explosion at Wellington Number Four Pit kills twenty-three; fire in Wellington Number Four Pit kills two.

Wellington Number Five shaft is opened.

1885
Victoria Theatre opens in October with Dunsmuir at centre stage.

Dunsmuir completes acquisition of an eleven-hectare lot for his new castle.

Lieutenant-Colonel Houghton fails to distinguish himself during the Riel Rebellion and is transferred to Montreal.

1886
Dunsmuir is re-elected to the legislature in July.

Sir John A. Macdonald drives the last spike of the E & N, which has been completed in record time.

Canada and the World

1884
Royal Commission on Chinese and Japanese Immigration hears testimony.

1885
Canadian Pacific Railway (CPR) is completed.

Chinese are denied the vote in federal elections and new arrivals must pay a fifty-dollar head tax.

Second Riel Rebellion occurs in the District of Saskatchewan, Northwest Territories.

1886
American Federation of Labor founded.

Laird of the Mines

DUNSMUIR AND HIS TIMES

1887
VCML explosion kills 148; the Chinese are blamed.

Provincial cabinet proposes Dunsmuir for a knighthood; Dunsmuir is named President of the Executive Council in August.

Building of Craigdarroch Castle begins.

Dunsmuir grants an interview to the *Portland News* in which he makes remarks some think to be treasonous.

1888
Dunsmuir begins to develop the mines at Cumberland.

Explosion in Wellington Number Five Pit in January kills sixty men; in the aftermath Dunsmuir agrees to exclude Chinese from his Wellington mines; he sends the Chinese to Cumberland to build the new company railway.

Dunsmuir receives Black Hand letters between August and October; the culprit is tried and jailed; *Victoria Times* says Dunsmuir carries the government in his breeches pocket; Dunsmuir sues the newspaper and wins.

CANADA AND THE WORLD

1887
Queen Victoria celebrates her Golden Jubilee.

Robert Dunsmuir

DUNSMUIR AND HIS TIMES

1889
The Knights of Labor threaten a strike in January; Dunsmuir locks his employees out; a delegation meets with him in Victoria; Dunsmuir threatens evictions; the men return to work; the miners at Cumberland strike.

Dunsmuir seeks advice from a spiritualist.

Dunsmuir, 63, takes sick on April 5; dies on April 12 not having signed his new will.

A public funeral for Dunsmuir occurs on April 16.

Reverend Hills visits Marion Dunsmuir Houghton in Montreal in June and hears about Dunsmuir's state of mind before his death.

Agnes Dunsmuir Harvey, 40, dies of typhoid fever in September followed by her husband five months later.

1890
Joan Dunsmuir and three unmarried daughters move into Craigdarroch Castle in September.

Wellington miners begin a strike for union recognition which lasts eighteen months and achieves nothing.

CANADA AND THE WORLD

1889
Washington becomes a state of the U.S.

1891
Sir John A. Macdonald dies.

Laird of the Mines

DUNSMUIR AND HIS TIMES

1892
Marion Dunsmuir Houghton dies at Craigdarroch.

1898
James is elected MLA for Comox.

Charles Frederick Houghton dies.

1899
James is sworn in as the Premier of B.C. in June.

Alexander Dunsmuir marries Josephine Wallace in December.

1900
Alexander, 47, dies in January an alcoholic.

1901
Elizabeth Dunsmuir Bryden, 54, dies.

1902
Alexander's step-daughter, Edna Hopper, sues James over Alexander's will; Joan Dunsmuir is party-plaintiff in the suit against her own son.

James resigns as the Premier.

1905
CPR buys the E & N Railway.

CANADA AND THE WORLD

1901
Queen Victoria dies and her son, King Edward VII, succeeds her.

1904
B.C. raises the Chinese head tax from $50 to $500.

Robert Dunsmuir

DUNSMUIR AND HIS TIMES

1906
James is appointed Lieutenant-Governor of B.C. in May; he serves until 1909.

1908
Joan Dunsmuir, 80, dies.

1910
Canadian Collieries (Dunsmuir) Ltd. buys all Dunsmuir possessions relating to coal mining in B.C. and California.

CANADA AND THE WORLD

1917
Electric lamps replace oil lamps in all the Vancouver Island mines.

1927
Canadian Collieries (Dunsmuir) Ltd. owns all coal mines on Vancouver Island.

1937
United Mine Workers of America sign union agreement with Canadian Collieries (Dunsmuir) Ltd.

Sources Consulted

Abbreviations:
British Columbia Archives – B.C. Archives.
British Columbia Genealogist – *BCG*.
Canadian Historical Review – *CHR*.
Craigdarroch Castle – CC.
Hudson's Bay Company Archives-Provincial Archives of Manitoba – HBCA-PAM.
Nanaimo District Museum – NDM.

Articles:
GALLACHER, Daniel T. "Dunsmuir, Robert," *Dictionary of Canadian Biography*, Vol. XI, 1881 to 1890. Toronto: University of Toronto Press.
———. "Men, Money, Machines. Studies Comparing Colliery Operations and Factors of Production in British Columbia's Coal Industry to 1891." UBC PhD Dissertation, 1979.
LAVIN, J. Anthony. "The Date and Place of Robert Dunsmuir's Birth," *BCG*, Vol. 10, No. 4, Winter 1981.
PORTER, Brian J. "Robert Dunsmuir – An Exercise in Genealogical Reconstruction," *BCG*, Vol. 10, No. 2, Summer 1981.

ROY, R.H. "In Aid of the Civil Power, 1877," *Canadian Army Journal*, 7:3, 1953.
SMITH, Allan. "The Myth of the Self-Made Man in English Canada, 1850-1914," *CHR*, 59 (1978).
VICKERS, Randolph Sydney, "George Robinson: Nanaimo Mine Agent," *The Beaver*, Autumn 1984.
"Victoria's Jay Gould," *The Daily Examiner*, April 25, 1887.

Books:
BOWEN, Lynne. *Boss Whistle*. Lantzville: Oolichan Books, 1982.
———. *Three Dollar Dreams*. Lantzville: Oolichan Books, 1987.
BOWSFIELD, Hartwell, ed. *Fort Victoria Letters 1846-1851*. Winnipeg: Hudson's Bay Record Society, 1979.
COLVILE, Eden. *Letters 1849-1852*. London: Hudson's Bay Record Society.
HELMCKEN, John Sebastian. *The Reminiscences of Doctor John Sebastian Helmcken*. Dorothy Blakey Smith, ed. Vancouver: University of British Columbia Press, 1975.
MARQUIS OF LORNE, K.T. *Canadian Pictures*. London: Richard Clay & Sons, 1885. CC.
REKSTEN, Terry. *Craigdarroch. The Story of Dunsmuir Castle*. Victoria: Orca Book Publishers, 1987.
———. *The Dunsmuir Saga*. Vancouver: Douglas & McIntyre, 1991.
Views of British Columbia and Alaska. Victoria: M.M. Waitt & Co. Booksellers. CC.

Correspondence:

British Columbia. Correspondence Outward, Inspector of Mines 1877-1900, B.C. Archives.

Houghton, Charles Frederick. Letterbook No. 1 from December 1, 1873 to May 30, 1877, B.C. Archives.

Hudson's Bay Company. Fort Victoria Correspondence to Hudson's Bay Company on Affairs of Vancouver Island Colony May 16, 1850-November 6, 1885, HBCA-PAM.

———. London Correspondence Inward, general, July-December 1848, HBCA-PAM.

———. London Correspondence Inward, general, Landale to Barclay and London correspondence outward, general, Barclay to Landale, June-December 1850, HBCA-PAM.

———. London Correspondence Inward, general, Robinson to Barclay and London correspondence outward Barclay to Robinson, March-May 1854, HBCA-PAM.

———. Nanaimo Letterbook, NDM.

Pearce, B.W. Correspondence with Robert Dunsmuir, 1871, B.C. Archives.

Diaries and Journals:

Bryden, John. Diary and Letterbook. 1878-1880, B.C. Archives.

Hills, Reverend George, B.C. Archives.

Hudson's Bay Company. Account Book of 1854.

Stuart, Captain Charles E. "Nanaimo Journal: August 1855-March 1857," NDM.

Government Records:

British Columbia. Inquisitions from April 1879 to July 1891. B.C. Archives.

———. Inquisition on Explosion May 3, 1887. B.C. Archives.

———. *Ministry of Mines Annual Reports*, 1874-1900. Victoria: Richard Wolfenden, Government Printer.

———. Sessional Papers, 1877-1901.

Canada. Report of the Adjutant-General for Military District No. 11, 1877. B.C. Archives.

———. *Report of the Royal Commission on Chinese and Japanese Immigration*. Ottawa: Queen's Printer, 1885.

Newspapers:
British Colonist
Daily Colonist
Nanaimo Free Press
Nanaimo Daily Free Press
Nanaimo Gazette

Index

Aboriginals. *See* First Nations people

Afterdamp. *See* Gas in mines

Alcohol use and abuse, 34, 35, 36, 43, 51, 70-71, 72, 74, 85, 103, 108, 114, 145
  See also Dunsmuir, Robert, and alcohol

Ardoon. *See* Dunsmuir, Robert, homes of

Banning. *See* Presbyterian Church

Beaver Harbour. *See* Fort Rupert

Begbie, Sir Matthew Baillie (judge), 5, 91-92, 136

Black Hand, 1-5, 135

Blasting, 44, 125, 131

Blowing out shots. *See* Explosions

Boxer, HMS, steam trials, 65-66, 73-74

Boys in the mines. *See* Labour, boy

Bryden, John (son-in-law), 55, 56, 106-07, 108, 109, 114, 121, 125, 133, 134, 137-38

Cages, 48, 49, 131

Canadian Pacific Railway (CPR), 113, 120

Chantrell's (saloon), 67-68, 70-71, 82, 93

Chinese, 108, 112
  as mine workers, 56, 57, 69, 77-78, 79, 105-07, 118-21, 132
  exclusion from the mines of, 104, 120-21, 125, 133-34, 138
  immigration of, 56-57, 119
  persecution of, 4, 57, 78, 119-120
  See also Labour, and the Chinese; Dunsmuir, Robert, and the Chinese

Chokedamp. *See* Gas in mines

Coal, finding of, 23, 24, 25-26, 27, 37, 51, 62
  See also Dunsmuir, Robert, as a prospector

Coal lands, 5, 119, 135

Coal loading, 12, 22, 23, 47-48, 65-66, 72
  See also Haulage

Coal miners, 11, 25, 32-33
  and their contracts, 11-12, 13, 26, 27, 34, 38
  and their wages, 12, 26, 54, 56, 57, 78, 80-81, 83, 84, 92, 95, 96, 101, 118, 120
  Scottish, 7, 11, 12-13
  Staffordshire, 33-36, 38, 57, 103
  See also First Nations people, as coal miners and loaders; Labour

Coalminers' Mutual Protective Society. *See* Unions

Coal mining, 32, 44, 49, 83
  government regulation of, 104, 119, 133
  in Scotland, 7, 13, 14
  See also Blasting; Cages; Lamps; Scales; Ventilation systems

Coal Mines Regulations Act of 1877. *See* Coal mining, government regulation of

Coal seams, 37, 43, 63, 65

Communications, 12, 78, 123

Company houses. *See* Housing, company

Craigdarroch Castle, 5, 6, 15, 121-22, 126-27, 135, 144

173

Robert Dunsmuir

Cumberland, B.C., 5, 133, 136, 137-38

Davie, Theodore (lawyer, future B.C. premier), 86, 94, 130, 136
Departure Bay, 53, 65, 72, 78, 84-85, 86, 87, 98, 100, 108, 111
Diggle, Wadham Neston (Dunsmuir's partner), 65, 69, 71, 86, 110
Douglas, James (later Sir), 34, 64, 97, 139
  as Chief HBC Factor, 12, 25-26, 27, 28, 34, 37, 38-39, 41, 42
Douglas Pit, 48-50, 54, 66
Dunsmuir, Agnes (daughter), 52, 61, 140
Dunsmuir, Alexander (son), 40, 52, 69, 71, 73, 82, 83, 108, 118, 127, 133, 134, 145
Dunsmuir, Annie Euphemia (daughter), 55, 140
Dunsmuir, Diggle & Company, 62, 84, 108
Dunsmuir, Elizabeth Hamilton (daughter), 9, 52, 55, 140
Dunsmuir, Emily Ellen (daughter), 52, 140
Dunsmuir, Henrietta Maude (daughter), 73, 138, 140
Dunsmuir, James (son), 19, 28-29, 40, 52, 69, 71, 73, 81-82, 83, 85, 86, 107-08, 127, 133, 145, 148
Dunsmuir, Jessie Sophie (daughter), 55
Dunsmuir, Joan Oliver White (wife), 52, 63, 123, 140, 144, 145
  domestic life of, 24, 34, 40-41, 110, 111
  early life in Scotland, 9-10, 109
  influence on her husband, 30, 41, 73, 126, 139
  pregnancies of, 9, 11, 14, 18-19, 27, 73
  and the promise, 15, 126, 144
Dunsmuir, Marion (daughter), 43, 52, 100-01, 140
Dunsmuir, Mary Jean (daughter), 52, 140
Dunsmuir, Robert, 27-28, 108-09, 111, 132, 133, 137
  and alcohol, 55, 67-68, 117-18, 124
  appearance of, 9, 85, 117, 141
  birth and childhood of, 4, 7-9
  and the Chinese, 105, 118, 120, 133, 138
  and the community, 52-53, 55, 73, 94-95, 109, 134, 140
  death and funeral of, 138-141, 144
  discovers and develops coal, 61-69
  emigration of, 4, 14-15
  as an employer, 42-45, 58, 134, 137
  homes of, 34, 43, 46, 55, 72-73, 100, 110, 111, 140-41
  influential friends of, 2, 86, 96, 97, 119, 126, 136
  investments of, 110, 119, 122, 127
  proposed knighthood of, 126, 128
  and labour, 49, 53, 69-70, 80-82, 83-84, 105, 121, 134
  and the law, 87, 88, 91-92, 93, 127-28, 136-37, 138
  as a manager, 4, 41-45, 49, 51, 53-54, 55, 56, 60, 105, 107-08
  marriage of. *See* Dunsmuir, Joan
  use of the military by, 87-88, 93, 95-100
  as a miner, 4, 9, 25, 26, 28, 36, 37-38, 53

174

and partners, 62-63, 68-69, 71, 110
as a politician, 109-10, 122-23, 126, 127-28, 132, 138
and the press, 53, 64, 70, 83-84, 88, 94-95, 96, 121, 127-28, 140
as a prospector, 26, 42, 61-65
Scottish ancestors of, 6-7
and unions, 83-85, 92, 105, 114, 118, 120, 132, 133-34, 137-38, 142, 146
will of, 139, 145
See also Black Hand; Craigdarroch Castle; Esquimalt and Nanaimo Railway

Elections, 32, 109-10, 122-23
Elliott, Andrew (B.C. premier and attorney general), 86, 92, 96, 97
Esquimalt and Nanaimo Land Grant, 5, 102, 114, 116, 119, 135-36
Esquimalt and Nanaimo Railway (E & N), 112-14, 121, 123-24, 127, 132, 133, 134
Evictions. *See* Housing, company
Explosions, 50, 90
in Nanaimo, 124-25
in Wellington, 106-07, 121, 131-33

Fairview. *See* Dunsmuir, Robert, homes of
Firedamp. *See* Gas in mines
Fires, 44, 71, 79-80, 104, 105-07
First Nations people, 10, 74
as coal miners and loaders, 12, 22, 23, 26, 32, 69
relationship with Europeans, 19, 22-23, 24
See also Kwakiutl; Nuwitti; Sne ney mux

Fort Rupert, Vancouver Island, 11-12, 15, 19, 20-28, 40, 140
Fort Vancouver, Oregon Territory, 17, 18-19
Fort Victoria, Vancouver Island, 3, 12, 27

Gas in mines, 50, 77, 105-07, 121, 125, 131-32
Gilmour, Boyd (uncle), 8-9, 13-14, 19, 21, 23-28, 34, 37, 41
Gilmour Jean Dunsmuir (aunt), 8, 12, 24, 27, 34
Gilmour party, 13, 17-19, 20-21, 25
Gold rushes, 12, 18, 25, 32, 36, 43, 56
Gough, Reuben (miner), 103, 107
Greenwell, John (miner), 132, 137-38
Griffin, Dr. Gustavus Hamilton. *See* Black Hand

Hamilton, Jimmie (Dunsmuir's friend), 61-62, 73-74
Harewood Coal Company, 51, 53-54
Harvey, James (son-in-law), 61, 69, 109
Haulage, 48-49, 65, 68, 71-72, 79
Head tax. See Chinese, immigration of
Helmcken, Dr. John (physician), 22-23, 50, 139
Hoggan, Alex (miner), 95-96, 101, 105
Houghton, Lt.-Col. Charles (son-in-law), 97-98, 100-01
Housing, 24, 33-35, 46, 70-71
company, 83, 87, 88, 91, 93-94, 95-96, 99-100, 118, 138
Hudson's Bay Company (HBC), 10, 31-33, 39, 44, 51, 149
London office of, 12, 14, 26

175

recruitment of miners, 11-13, 33
See also Fort Rupert, Fort Vancouver, Fort Victoria

Indians. *See* First Nations people
Inspectors, 86, 104, 106

Knights of Labor, 118-19, 120, 132, 133-34, 137
Kwakiutl, 19, 21-22, 23, 26 32, 39-40

Labour, 32-33, 35, 42-43, 49, 69-70, 105, 121, 123
and the Chinese, 56-57, 77, 96, 105, 118
boy, 103-04
desertions of, 12, 35-36, 38, 42
grievances of, 81, 83, 88, 137-38
See also Coal miners; Strikes; Unions
Lamps, 35, 168
fish oil (naked flame), 49-50, 90, 106-07, 121, 125
safety, 90, 106
Landale, David (HBC recruiter), 11, 12-14, 15, 21
Law suits. *See* Trials
Locomotives, 47-49, 71-72
Lorne, Marquis of (Governor General), 4, 67, 111-13

Macdonald, Sir John A. (Prime Minister of Canada), 113, 123-24
Mail. *See* Communications
Markets, 41, 42, 53, 54, 80
Maude (private railway car), 123, 134
Maude (side-wheel steamer), 92, 98, 100, 101
Meakin, John (miner), 36, 51-52

Methane gas (firedamp). *See* Gas in mines
Militia. *See* Dunsmuir, Robert, use of the military by
Miners. *See* Coal miners
Mines. *See* Douglas Pit; Wellington Colliery, mines
Muir, John and family, 11-12, 13, 22, 23, 25, 27, 41, 62
Murders, 13-14, 19-20, 21, 74

Nanaimo, B.C., 31, 37, 43, 46, 51-52, 55, 78, 111-12, 124
harbour, 27, 37, 39, 46, 50, 54, 124
bastion, 33-34, 36, 38, 39-40, 46
Natives. *See* First Nations people
Newspapers, 78, 80, 87, 88, 141
See also Dunsmuir, Robert, and the press
Nuwitti, 19-20, 21

Pekin, 15, 17-18
Police, 92-93, 101, 118
Presbyterian Church, 6, 7-8, 9-10, 141
Prospecting. *See* Coal, finding of

Railways, 53-54, 133, 136
See also Canadian Pacific; Esquimalt and Nanaimo; Wellington Colliery, railway
Robinson, George (mine manager), 35-36, 40, 41-44, 62
Rocket, HMS, 87-88, 93
Royal Navy, 10-11, 19-20, 27, 51, 54, 65-66, 69, 74

Safety, 105, 125
See also Inspectors; Coal mines, government regulation of
San Francisco, California, 41, 65, 80, 108

Laird of the Mines

Scales, 48, 81-82, 83, 88
Settlement Act of 1884, 114, 119, 122-23
Sheriff Harris, 93-94, 95-96, 99-100
Ships, 10, 19, 20, 34, 64, 121
   coal (colliers), 47-48, 50, 53, 110, 118
   See also Boxer, *Maude*, *Pekin*, *Rocket*
Smallbones family, 67-68, 73-74
Sne ney mux, 27, 32, 39-40, 47
Spalding, Captain Warren (stipendiary magistrate), 86-87, 91, 93, 96, 97, 98, 99
Special police. *See* Police
Staffordshire miners. *See* Coal miners, from Staffordshire
Strikebreakers, 57, 83-85, 92-93, 95, 101, 118
Strikes, 11-12, 28, 56, 114, 146
   Wellington 1876, 80-82
   Wellington 1877, 83-101, 104, 113, 118
   Wellington 1883, 118, 120
   Wellington 1889, 137
Stuart, Captain Charles (Officer-in-Charge, Nanaimo), 31-32, 36, 38, 42, 43

Telegraph and telephone. *See* Communications
Trials, 4-5, 101
   See also Dunsmuir, Robert, and the law

Unions, 81, 83-85, 88, 95, 99, 118
   legislation concerning, 57-58, 149
   See also Knights of Labor; Dunsmuir, Robert, and unions

Vancouver Coal Mining and Land Company (VCML), 44-45, 47-50, 53-54, 56, 69, 107, 111, 124-25, 133
Ventilation systems, 50, 77, 79, 105, 121, 125, 132
Victoria, B.C., 1-6, 101, 110, 121-22, 134-35, 140-41
Voter qualifications. *See* Elections
Voyages, 11, 13, 17-18, 35

Wages. *See* Coal miners, and their wages
Walkem, George (lawyer, B.C. premier), 86, 92, 93, 113
Walker, Edward (friend), 24-25, 27, 28, 36-37, 38, 140
Weighmen. *See* Scales
Wellington, B.C., 70-71, 74, 88, 93, 95-96, 98, 99-100, 112, 146
Wellington Colliery, 120, 146
   mines, 76, 77, 79, 105-07, 121, 125, 140
   offices of, 72, 79, 95, 96, 101, 112
   railway, 71-72, 99, 110, 116
Wellington Inn. *See* Chantrell's
Wellington seam. *See* Coal seams
Wharves. *See* Coal loading.

*Printed in June 2004
at Marc Veilleux imprimeur,
Boucherville (Québec), Canada.*